Ernst Schering Research Foundation Workshop 20
Cellular Therapy

Springer-Verlag Berlin Heidelberg GmbH

Ernst Schering Research Foundation
Workshop 20

Cellular Therapy

H. Wekerle, H. Graf, J.D. Turner
Editors

With 24 Figures and 19 Tables

 Springer

Series Editors: G. Stock and U.-F. Habenicht

ISSN 0947-6075

ISBN 978-3-662-03511-5   ISBN 978-3-662-03509-2 (eBook)
DOI 10.1007/978-3-662-03509-2

CIP data applied for

Die Deutsche Bibliothek – CIP-Einheitsaufnahme
Schering-Forschungsgesellschaft <Berlin>: Ernst Schering Research Foundation Workshop. - Berlin; Heidelberg; New York; Barcelona; Budapest; Hong Kong; London; Milan; Paris; Santa Clara; Singapore; Tokyo: Springer.
ISSN 0947-6075
20. Cellular therapy. - 1997
Cellular therapy / H. Wekerle ... ed. - Berlin; Heidelberg; New York; Barcelona; Budapest; Hong Kong; London; Milan; Paris; Santa Clara; Singapore; Tokyo: Springer, 1997
(Ernst Schering Research Foundation Workshop; 20)

© Springer-Verlag Berlin Heidelberg 1997
Originally published by Springer-Verlag Berlin Heidelberg New York in 1997.
Softcover reprint of the hardcover 1st edition 1997

Typesetting: Data conversion by Springer-Verlag
SPIN: 10534108   13/3135–5 4 3 2 1 0 – Printed on acid-free paper

Preface

The use of cells for the treatment of a variety of diseases is no longer a dream. Today, blood transfusion, bone marrow transplantation, the use of ex vivo cultured skin in wound healing, and peripheral stem cell transplantation, including the ex vivo expansion of hematopoietic stem cells after high-dose chemo/radiation therapy, are routine. This high standard of knowledge and skills in cell transplantation might also result in tackling hitherto untreatable diseases.

Organ transplantation is presently the only life-saving treatment for a variety of conditions. Important findings in cell and molecular biology, the identification of hematopoietic, mesenchymal and neuronal stem cells, together with breakthroughs in the methodology for isolating, purifying, expanding, and storing human cells could make cellular therapy an alternative to organ transplantation in certain diseases within the next decade. Placental blood may be the source of choice in isolating naive progenitor cells for allogeneic transplantation.

Immunotherapy is the most hopeful strategy to date for the treatment of tumors resistant to chemotherapy, radiation therapy, or hormone therapy. It includes the use of tumor-infiltrating lymphocytes, ex vivo activated memory T lymphocytes, and cell-based vaccines.

The transplantation of glial cells is a hopeful new strategy in demyelinating diseases such as multiple sclerosis, and encephalitogenic T cells have been shown to be capable of crossing the blood–brain barrier, secreting their transgene products within the area of interest. They could therefore be used as autologous cellular vehicles for gene therapy of CNS diseases.

The participants of the workshop

To meet these therapeutic needs, the major forces and obstacles for a meaningful translation of such novel approaches into the development of cellular therapeutic products must be identified. Multi-disciplinary cooperation will be a must for success in developing cellular therapy, and there will be a considerable overlap to gene therapy – under today's more realistic and more cautious approaches, transplantable human cells will be important vehicles for somatic gene therapy.

In addition to combined efforts in research and development, strategic alliances have been formed between clinics, biotech companies, academic institutions, and pharmaceutical companies. These alliances will have to build a very intelligent infrastructure in order to fulfill the highly complex needs of cell transplantation in view of an effective cellular therapy for the benefit of the patients.

I hope that the workshop held in Berlin, Germany, in August 1996, the proceedings of which have been published in the present volume, will help to develop a better understanding – and even more importantly – better cooperation in this challenging new field.

G. Stock

Table of Contents

List of Editors and Contributors

Editors

H. Graf
Schering AG Berlin, c/o Metron GmbH, Elsässer Straße 2n,
79110 Freiburg i.Br., Germany

J.D. Turner
Schering AG, CNS Research, Müllerstraße 178, 13342 Berlin, Germany

H. Wekerle
Max-Planck-Institute for Psychiatry, 82152 Martinsried, Germany

Contributors

R. Andreesen
University of Regensburg, Department of Hematology and Oncology,
93042 Regensburg, Germany

R.G. Bretzel
Justus-Liebig-Universität, III. Medizinische Klinik und Poliklinik,
Rodthohl 6, 35385 Gießen, Germany

K. Federlin
Justus-Liebig-Universität, III. Medizinische Klinik und Poliklinik,
Rodthohl 6, 35385 Gießen, Germany

R.J.M. Franklin
University of Cambridge, Department of Clinical Veterinary Medicine
and MRC Cambridge Centre for Brain Repair, Madingley Road,
Cambridge CB3 0ES, UK

U. Haberkorn
German Cancer Research Center, Department of Oncological Diagnostics and
Therapy, Im Neuenheimer Feld 280, 69120 Heidelberg, Germany

A. Hartmann
Klinikum der Philipps-Universität Marburg, Zentrum für Nervenheilkunde,
Klinik für Neurologie mit Poliklinik, Rudolf-Bultmann-Straße 8,
35039 Marburg

B. Hennemann
University of Regensburg, Department of Hematology and Oncology,
93042 Regensburg, Germany

R. Henschler
University Medical Center, Department of Hematology and Oncology,
Hugstetter Straße 55, 79106 Freiburg i.Br., Germany

B.J. Hering
Justus-Liebig-Universität, III. Medizinische Klinik und Poliklinik,
Rodthohl 6, 35385 Gießen, Germany

A. Konur
University of Regensburg, Department of Hematology and Oncology,
93042 Regensburg, Germany

S.W. Krause
University of Regensburg, Department of Hematology and Oncology,
93042 Regensburg, Germany

S. Kressenstein
Ludwig-Maximilians-University, Institut für Immunologie, Goethestraße 31,
80336 München, Germany

M. Kreutz
University of Regensburg, Department of Hematology and Oncology,
93042 Regensburg, Germany

A. Kupsch
Humboldt-University, Department of Radiology, Berlin, Germany

W. Lange
University Medical Center, Department of Hematology and Oncology,
Hugstetter Straße 55, 79106 Freiburg i.Br., Germany

A. Lindemann
University Medical Center, Department of Hematology and Oncology,
Hugstetter Straße 55, 79106 Freiburg i.Br., Germany

B. Maget
Ludwig-Maximilians-University, Institut für Immunologie, Goethestraße 31,
80336 München, Germany

R. Mertelsmann
University Medical Center, Department of Hematology and Oncology,
Hugstetter Straße 55, 79106 Freiburg i.Br., Germany

D. Möbest
University Medical Center, Department of Hematology and Oncology,
Hugstetter Straße 55, 79106 Freiburg i.Br., Germany

H. Neumann
Max-Planck-Institute for Psychiatry, 82152 Martinsried, Germany

E. Nößner
Ludwig-Maximilians-University, Institut für Immunologie, Goethestraße 31,
80336 München, Germany

R. Oberneder
Ludwig-Maximilians-University, Urologische Klinik und Poliklinik,
Goethestraße 31, 80336 München, Germany

W.H. Oertel
Klinikum der Philipps-Universität Marburg, Zentrum für Nervenheilkunde,
Klinik für Neurologie mit Poliklinik, Rudolf-Bultmann-Straße 8,
35033 Marburg

K. Pantel
Ludwig-Maximilians-University, Institut für Immunologie, Goethestraße 31,
80336 München, Germany

O. Pogarell
Klinikum der Philipps-Universität Marburg, Zentrum für Nervenheilkunde,
Klinik für Neurologie mit Poliklinik, Rudolf-Bultmann-Straße 8,
35039 Marburg

D.J. Prockop
Center for Gene Therapy, Allegheny University of the Health Sciences,
MCP- Hahnemann School of Medicine, 245 North 15 Street, Mail Stop 421,
Philadelphia, PA 19102, USA

F.M. Rosenthal
University Medical Center, Department of Hematology and Oncology,
Hugstetter Straße 55, 79106 Freiburg i.Br., Germany

D.J. Schendel
Ludwig-Maximilians-University, Institut für Immunologie, Goethestraße 31,
80336 München, Germany

A. Spyridonidis
University Medical Center, Department of Hematology and Oncology,
Hugstetter Straße 55, 79106 Freiburg i.Br., Germany

C.F. Waller
University Medical Center, Department of Hematology and Oncology,
Hugstetter Straße 55, 79106 Freiburg i.Br., Germany

R.M. Weber-Nordt
University Medical Center, Department of Hematology and Oncology,
Hugstetter Straße 55, 79106 Freiburg i.Br., Germany

H. Wekerle
Max-Planck-Institute for Psychiatry, 82152 Martinsried, Germany

J. Winkler
University Medical Center, Department of Hematology and Oncology,
Hugstetter Straße 55, 79106 Freiburg i.Br., Germany

1 Somatic Gene Therapy in Cancer

F.M. Rosenthal, C.F. Waller, A. Lindemann, and R. Mertelsmann

1.1 Background

Cancer is rapidly becoming the leading medical cause of death in Western society. Currently, the most important treatment modalities in clinical oncology are surgery, radiotherapy, and chemotherapy with cytostatic or cytotoxic drugs. Even though important progress has been

Table 1. Gene transfer techniques

---

1.   Non-viral methods
1.1 Physical methods
     – Electroportion
     – Microinjection
     – Particle bombardment
     – Lipofection
1.2 Chemical methods
     – Calciumphosphate co-precipitation
1.3 Receptor-mediated transfer
     – DNA/protein complexes
     – DNA/virus complexes

2.   Viral vectors
2.1 Retroviruses
2.2 Adenoviruses
2.3 Adeno-associated viruses
2.4 Herpes simplex viruses
2.5 Other: Eppstein-Barr viruses, HIV, Vaccina viruses, Polio viruses,
     SV40 viruses

3.   Bacterial vectors
     Salmonella, Listeria

---

made using these therapeutic modalities alone or in combination, treatment results have been stagnating in recent years in many types of cancer, especially in those which occur more frequently, such as lung cancer or cancer of the colon, in spite of intensive clinical research.

Investigation of cancer cells with molecular techniques has provided new insights into the pathogenesis and pathophysiology of cancer. It has been shown that tumors arise secondary to genetic changes in cells, which result in abnormal proliferation of the mutated cell clone. A better understanding of tumor pathogenesis and progress in molecular biology, cell biology, and immunology has now provided the foundation for the development of new therapeutic strategies for patients with cancer.

Gene therapy is the introduction of a gene into a cell which then produces the desired gene product and thus corrects a genetic defect or acquires a new function (Anderson 1984; Friedman 1992). Somatic

Table 2. Gene therapeutic strategies in cancer

---

1. Genetic marking
   - Effector cells of the immune system
   - Tumor cell contamination in hematopoietic stem cell preparation
   - Hematopoietic stem cells
2. Gene transfer into effector cells of the immune system
   - Cytokines
   - Chimeric antigen receptors
3. Gene transfer into cancer cells
3.1 Correction of the genetic defect
   - Anti-oncogene wild type (e.g., p53)
   - Oncogene antisense (e.g., K-ras)
   - Intracellular single chain antibodies
3.2 Augmentation of antitumor immune response
   - Cytokines
   - Cytokine receptors
   - MHC molecules
   - Antigens (allogeneic, superantigen, tumor specific)
   - Costimulatory molecules
3.3 Transfer of suicide genes
   - Herpes simplex virus thymidine kinase
   - Cytosine deaminase
4. Gene transfer into antigen-presenting cells and bystander cells
   - Antigens
   - Costimulatory molecules
   - Cytokines
5. Chemo- and radioprotection of non-neoplastic tissues
6. Systemic production of therapeutic molecules
   by genetically modified cells in vivo

---

gene therapy has to be distinguished from germ line therapy, in which the therapeutic gene is introduced in sperm, ova, or embryonal stem cells. The purpose of germ line gene therapy is to prevent the transmission of defective genes to subsequent generations. However, the use of germ line therapy in men is far from being a standard clinical practice, not only because technology is not sufficient but also because there is insufficient knowledge to evaluate the risks to future generations and thus a number of ethical and social problems remain unsolved.

Gene transfer into cells of therapeutic interest can be achieved by a variety of strategies. Gene transfer techniques can be principally divided into nonviral and viral methods (Table 1) (Flynn et al. 1990; Pan et al. 1995; Felgner et al. 1987; Proctor 1992; Yang et al. 1990; Zenke et al. 1990; Eglitits et al. 1985; Chasse et al. 1989; Kotin et al. 1990; Kozarsky and Wilson 1993; Breakefield and DeLuca 1991; Moss 1991; Shimada et al. 1991). Viral vectors are modified to be nonpathogenic and replication defective. Current methods of gene transfer either lead to transient expression due to episomal (nonintegrated) localization of the transfected gene, which gets lost in the following cell divisions as it does not replicate with the DNA in the nucleus, or they lead to random, but stable integration of the plasmid into the genome. In this case, the transfected gene is passed on to daughter cells after cellular division.

Several gene therapeutic strategies for patients with cancer have already made their way into early clinical trials (Rosenthal and Mertelsmann 1997). A review of these approaches is presented below and outlined in Table 2.

1.2 Gene Marking

The genetic marking of cells allows the trafficking, survival and functional properties of the marked cell to be investigated following adoptive transfer into the patient (Fox et al. 1989). The marker gene can be demonstrated later on in isolated cells by sensitive detection methods such as the polymerase chain reaction (PCR). Since the genetic marker does not decay or become diluted as cells replicate in vivo, these cells can also be followed up long term in vivo. This strategy has no direct therapeutic goal, but several aspects of the pathogenesis and biology of a disease as well as the physiological aspects of hematopoiesis or the immune system can be studied in vivo.

In 1989, the first clinical gene transfer protocol was initiated at the National Institutes of Health in the USA. The patient received autologous tumor infiltrating lymphocytes (TIL) transducted with the bacterial cDNA encoding resistance to the neomycin analog G418 (neomycin phosphotransferase). Apart from TIL other effector cells of the immune system such as cytotoxic T lymphocytes (CTL) can also be genetically marked in a similar way. This approach is also currently being used in several clinical trials of high-dose chemotherapy followed by autologous bone marrow transplantation (Brenner et al. 1993). Autologous stem cells from the bone marrow or the peripheral blood are genetically marked in vitro and reinfused after high-dose chemotherapy. At the time of relapse, the genetic tagging should help to decide on the origin of the relapse. If genetically marked clonal tumor cells can be demonstrated, this indicates contamination of the transplanted cell population with residual tumor cells and therefore a failing of the purging techniques performed ex vivo. In one clinical study, it has been shown that bone marrow from patients that were regarded as tumor free by PCR before transplantation still contained tumor cells that were able to contribute to the recurrence of the tumor (Rill et al. 1994).

In recent years, peripheral stem cell transplantation after high-dose chemotherapy has gained importance in the treatment of several malignancies. As very long term follow-up (>2 years) of transplanted patients has not been conducted so far, final evaluation of the complete and permanent reconstitution of all hematopoietic cell lineages by this treatment has not been possible. Gene marking of blood and marrow hematopoietic progenitor cells before transplantation should help to answer this question. Recently, a preliminary report on a human autologous transplantation study in patients with multiple myeloma or breast cancer using retrovirally transduced bone marrow and peripheral blood CD34+-enriched cells was published (Dunbar et al. 1995). The results suggest that the progeny of mobilized peripheral blood cells as well as bone marrow cells can contribute to long-term engraftment (three patients were studied for at least 18 months) of multiple lineages. Similar observations demonstrating long-term engraftment of peripheral blood hematopoetic precursor cells (HPCs) have been made in the setting of allogeneic transplantation (J. Finke, personal communication).

1.3 Gene Transfer into Effector Cells of the Immune System

1.3.1 Cytokines

The modification of anticancer cytotoxic effector cells of the immune system is another promising treatment approach. Adoptive immune therapy with TIL and interleukin 2 (IL-2) can occasionally induce regression of metastasized cancer in certain patients. Early clinical studies with genetically marked TIL in patients with malignant mela- noma have shown that these TIL can localize to the tumor site. Thus, TIL may provide a vehicle for the delivery of therapeutic genes to the tumor. By transfer of certain cytokine genes, such as that for tumor necrosis factor-α (TNF-α), it is hoped that the therapeutic activity of these effector cells of the immune system can be augmented, protected, or redirected (Hwu et al. 1993).

1.3.2 Chimeric Antigen Receptors

T cells are able to detect single amino acid changes in proteins. Thus, mutations and rearrangements in cellular proteins – which are a hall- mark of tumor development – can theoretically serve as target for specific T lymphocytes. Therapeutic efficacy of specific CTL has been demonstrated in immune-compromised patients after bone marrow transplantation by administration of Epstein-Barr virus (EBV)-specific CTL, where adoptive transfer leads to regression of EBV-associated lymphoma (Papadopoulos et al. 1994; Smith et al. 1995). To therapeuti- cally explore the specific cytotoxicity of CTL, either specific CTL should be generated in the tumor patient in vivo or these specific T cells should be induced in vitro and then reinfused into the patient. While it is clear that specific T cells are potentially powerful therapeutic agents, it has been difficult to generate cytotoxic T cells with specificity for tumor-associated antigens (Riddell et al. 1992). Gene transfer technolo- gies now offer the possibility to genetically engineer cytotoxic effector cells, for example by transfer of chimeric antigen receptors into T cells (Gross and Eshhar 1992; Stancovski et al. 1993). The extracellular component or the chimeric antigen receptor consists of the variable region of an immunoglobulin molecule with specificity for a tumor-as- sociated antigen. The intracellular domain is combined with the trans- membrane and the cytoplasmic, signal-transducing sequence of the T cell receptor molecule. Thus T cell cytolytic activity is redirected by the specificity of the immunoglobulin gene and tumor-specific antibod- ies are changed into tumor-recognizing receptors on T cells (Moritz et al. 1994). It has been shown to thus bypass major histocompatibility complex (MHC) restriction of T cells. A refinement of this technology and preclinical studies, for example, on the necessity of costimulatory signals to maintain cytotoxic activity of the T cells, are needed before this technology can be transferred into the clinic.

1.4 Gene Transfer into Cancer Cells

1.4.1 Correction of the Genetic Defect

The modification of cancer cells in an effort to correct tumor-specific genetic alterations is, at first sight, the most logical step. By site-specific recombination, the defective gene(s) and the defective gene function(s) could be replaced by inserting a healthy gene. Another approach is the use of antisense strategies in order to suppress the expression of a gene which could be relevant for the malignant phenotype. In various in vitro models, this approach has demonstrated normalization of the transformed phenotype by correcting the genetic defect. Clinically, the central problem with this approach is the necessity to transfer the therapeutic gene into every cancer cell: The transfection efficiency of current gene transfer methods is not sufficient to achieve that, nor is it possible to restrictively target genes to certain cell populations in vivo. However, both in vitro and in several animal models it has been shown that is might be possible to eradicate malignant cells by treatment with antisense-based strategies. In chronic myelogenous leukemia (CML), the malignant clone has been shown to express the fusion gene bcr/abl in over 90% of patients (Gale et al. 1993). This fusion gene is a required factor for the proliferation of leukemic cells in vitro and in vivo and therefore offers an ideal target for antisense therapy. Several investigators have successfully applied ribozymes, catalytic RNA molecules, to target bcr/abl, resulting in the cleavage of the transcript (Lange et al. 1993; Snyder et al. 1993). Another novel approach might be the use of "2–5A-antisense" technology. The 2–5A system is a highly regulated pathway for controlling RNA stability. It has been possible to clone 2–5A-dependent RNAse, an enzyme which is present in most mammalian cells. Activation of this 2–5A-dependent RNAse results in the cleavage of single-stranded RNA. By combining this host–defense mechanism – which is able to cleave viral as well as cellular RNA with antisense technology – it is possible to target specific mRNA sequences. An antisense sequence complementary to the target is chemically linked to 2–5A, the activator of 2–5A-dependent RNAse. The 2–5A part of the chimera binds and activates the RNAse present in cells while the antisense moiety binds to the complementary sequence in the target mRNA, thus causing the selective destruction of the mRNA target. This approach has been successfully applied both in a cell-free system and in intact cells for another mRNA target (Torrence et al. 1993; Maran et al. 1994).

1.4.2 Augmentation of Antitumor Immune Response

The rationale for using immune-stimulating strategies in cancer therapy is based upon the knowledge that tumor-specific antigens can be presented in association with MHC class I or II molecules on the cell surface to cytotoxic or helper T cells. The minimal requirements for a tumor cell to be recognized by a T cell-mediated immune response are the expression of MHC molecules and the presence of an antigen that can be recognized. Costimulatory or accessory molecules are membrane-bound molecules on T cells and APC that facilitate T cell recognition and activation and therefore are also important. As tumor-specific cytotoxic T cells can only be rarely demonstrated in humans and are even more rarely associated with regression of advanced tumor, it has to be assumed that tumor-specific antigens are only insufficiently recognized by the immune system or immune-suppressing mechanisms are induced by the tumor. The specific reasons, however, why the immune system is ineffective in preventing growth of cancer cells are not completely understood. Several explanations have been postulated, including the lack of tumor-specific antigens in tumor cells, a lack of MHC molecules, deficiency in antigen processing or presentation, insufficient expression of costimulatory molecules as well as secretion of immune suppressive products by the tumor. Approaches to activate effector cells of the immune system by systemic administration of immune-stimulatory cytokines have only demonstrated limited efficacy in humans. Gene therapeutic approaches that focus on the transfection of genes encoding cytokines, costimulatory signals, MHC molecules, or allogeneic antigens into tumor cells are being pursued by several laboratories and have already prompted several clinical trials (Gansbacher et al. 1993; Pardoll 1992; Rosenthal et al. 1994a; Tepper and Mule 1994).

1.4.3 Cytokines

Cytokines can influence the immune response on two different levels. They can modulate the afferent arm of the immune response by direct action on target cells (e.g., enhancing tumor cell immunogenicity by upregulating molecules involved in antigen processing and presentation; activation of antigen presenting cells, APC) and they can modulate the efferent arm of the immune response by activating effector cells of the immune system. Systemic administration of immunomodulatory cytokines like IL-2 or interferon (IFN)-γ in cancer patients has occasionally resulted in complete tumor remissions. These remissions, however, are rare, and toxicities associated with these treatments are often substantial.

The introduction of cytokine genes into tumor cells resulting in the local expression of cytokines in close proximity to target and effector cells has proven successful in inducing antitumor immune responses in several animal models without observable systemic toxicity. Two general questions have been addressed in these studies: First, what local effect does cytokine secretion have on primary tumor growth of *unirradiated* cytokine gene-transfected cells in vivo, and which immunological mechanisms lead to tumor rejection in this setting, and can immunological memory be induced; secondly, what is the potential of *irradiated* cytokine gene-modified tumor cells to induce systemic tumor-specific immunity and immunological memory. With both approaches the effects on the growth of preestablished unmodified tumor cells were analyzed. Depending on the cytokine and the tumor model studied, different effector functions of the immune system were activated. The level of cytokine expression, number of injected tumor cells, location of immunization and challenge, and the mouse strain used are critical parameters. To date a wide variety of cytokines have been studied in many different tumor models.

Analyzing the potency of different cytokines to reject unirradiated cytokine gene transfected tumor cells, cytokines that do enhace (IL-1, -2, -4, -6, -7, IFN-α_1, -γ, TNF-α, granulocyte-colony stimulating factor,

G-CSF), do not alter (IL-5, -10 and macrophage-colony stimulating factor, M-CSF), or even decrease tumor immunogenicity (transforming growth factor, TGF-β) have been identified (Blankenstein et al. 1991: Gansbacker et al. 1990a,b; Golumbek et al. 1991; Hock et al. 1193a,b; Rosenthal et al. 1994b; Saito et al. 1994; Vieweg et al. 1994; Sun et al. 1995; Allione et al. 1994). Histological and immunohistological examinations of the site of tumor inocculation as well as in vivo depletion of effector cell subpopulations or experiments in effector cell-deficient mouse strains (nude, SCID, or beige mice) have revealed those cells mediating or contributing to primary tumor regression. Depending on the cytokine used, cellular infiltrates consisted predominantly of unspecific inflammatory cells such as macrophages, eosinophils, neutrophils, and natural killer (NK) cells or of CD4$^+$ or CD8$^+$ T lymphocytes. G-CSF secretion by murine colon carcinoma cells induces rejection of transfected tumor cells by a T cell-independent mechanism, i.e., by neutrophils, and does not lead to systemic imunity. However, even if primary rejection is not mediated by T cells, the establishment of a long-term protective immune response requires the presence of CD8$^+$ T cells and in some models also of CD4$^+$ T cells.

Whereas in most animal models it has not been possible to show rejection of preestablished disease, in some tumor systems this has been possible. IL-2, -4, -6, IFN-α$_1$ and GM-CSF have been shown to be effective in the treatment of some preestablished murine tumors of different histology (Tepper and Mule 1994; Dranoff et al. 1993; Connor et al. 1993).

As another way of obtaining sufficient viable cells in the clinical setting, the use of allogeneic tumor cell lines whose HLA class I antigens are identical with the patient has been proposed. This approach is based on the notion that certain HLA class I molecules present so called "shared tumor antigens" and an immune response directed against the vaccine would also recognize the autologous tumor (Gansbacher et al. 1992a,b).

1.4.4 Antigens

Another strategy for stimulating an antitumor immune response by genetic modification is based on the introduction of a highly immunogenic allogeneic MHC class I molecule (Hui et al. 1989). Nabel and coworker have used direct allogeneic MHC class I gene transfer to modify established tumors in mice. They have been able to demonstrate that introduction of this strongly immunogenic antigen into progressively growing tumors led to the generation of a T cell-mediated immune response against both the allogeneic MHC gene on transfected tumor cells as well as against previously unrecognized antigens on unmodified tumor cells. The induction of CTL resulted in partial tumor regression and in some cases in complete remissions in mice. While the complexity of this response is only partially understood, it was postulated that the expression of the highly immunogenic foreign transplantation antigen created an appropriate cytokine milieu at the tumor site, inducing an inflammatory response and a cellular infiltrate which led to the observed tumor-directed, tumor-specific cytotoxic T cell response. Based on these studies, a phase I clinical trial in patients with stage IV cutaneous melanoma was initiated. Liposome complexes containing a eukaryotic expression vector plasmid with the human HLA-B7 cDNA were directly injected into subcutaneous tumor nodules of HLA-B7-negative patients (Nabel et al. 1993). Patients were treated without complications and no antibodies to DNA were detected in any patient. The recombinant HLA-B7 protein was detected in tumor biopsy tissues from all patients but not in the peripheral blood of any of them. One patient showed regression of the injected tumor nodules, accompanied by regression at a distant tumor site. In two patients in whom tumor cell lines could be established, tumor-specific CTL were detected among the population of lymphocytes in the blood and lymphocytes infiltrating the tumor. In one patient an increase in anti-HLA-B7-specific CTL precursors could be observed. These results have demonstrated the feasibility, safety, and therapeutic potential of direct transfer of a foreign histocompatibility gene into human tumors.

1.4.5 Costimulatory Molecules

As many tumor cell types do not effectively express these important components of the immune response, their potential to act as stimulators or targets of a cell-mediated response is decreased. In several animal studies, the critical role of costimulatory molecules, such as the B7 surface molecule, in the generation of antitumor immune responses has been demonstrated (Yang et al. 1995). Immunogenic tumors that were not rejected by the immune system were rendered susceptible to T cell-mediated rejection by transfection of the B7 costimulatory molecule (Chen et al. 1994; Townsend and Allison 1993). A phase I vaccination trial of B7-transfected allogeneic melanoma cells has recently been initiated in melanoma patients in the United States. As other costimulatory molecules have been identified and characterized, it is likely that these will also play a role as future tumor vaccines, either alone or in combination with other immune stimulatory genes.

1.4.6 Transfer of Suicide Genes

Another promising approach in gene therapy for cancer is the insertion of genes coding for prodrug activating enzymes (suicide genes) into cancer cells. The herpes simplex virus thymidine kinase (HSV-tk) converts the nontoxic nucleoside analog ganciclovir (GCV) into a monophosphate form which can be converted into a triphosphate form by endogenous mammalian enzymes and that acts as a chain terminator in DNA synthesis. Cytosine deaminase (CD) is another suicide gene-encoded enzyme that has been used to confer selective cytotoxic sensitivity to transfected cells in vitro and in vivo. CD is an enzyme expressed in some bacteria and fungi that, apart from deaminating cytosine to uracil, also deaminate the relatively nontoxic 5-fluorocytosine (5-FC) to the highly toxic 5-fluorouracil (5-FU). Transfection of the HSV-tk or the CD gene into tumor cells renders them selectively sensitive to the appropriate prodrug in vitro and also to a certain extent in vivo (Mullen et al. 1994). Nontransfected normal mammalian cells that do not contain the suicide gene are relatively resistant to the prodrug. For this therapy to be effective in vivo, tumor cells of the tumor-bearing host would either have to be selectively transfected or the gene would have to be preferentially expressed in tumor cells. One way to achieve specific expression in cancer cells would be to use tumor-specific or tissue-specific promoters (Vile and Hart 1993a,b; Kuriyama et al. 1991).

The intratumoral injection of a retroviral producer cell line might be an alternative for the selective expression of a suicide gene in tumor cells (Culver et al. 1992). As retroviruses only integrate in replicating cells, this approach is especially suitable for the treatment of solid tumors that grow rapidly and invade normal tissue and that consist mainly of quiescent nonproliferating cells, such as in the brain (Miller et al. 1990). As opposed to neurons in the central nervous system, in proliferating brain tumors only tumor cells or endothelial cells of blood-supplying vessels would be dividing. In the 9L-gliosarcoma model, rats were implanted into the frontal lobe with tumor cells and were treated with an intratumoral stereotactic injection of murine retroviral HSV-tk producer cells 7 days later(Ram et al. 1993). In 80% of the animals, complete resolution of the tumor was observed after 14 days of treatment with GCV. Some 50%–60% of the rats showed long-term survival, despite the fact that only a small percentage of the tumor cells showed integration of the retroviral vector. The efficacy of this treatment appears to be due to a so called "bystander-effect" (Freeman et al. 1993). It is thought to be related to several factors, including the transfer of toxic GCV metabolites via intracellular gap junctions, the phagocytosis of apoptotic vesicles from nearby, unmodified tumor cells, a reduced blood flow to the tumor due to death of HSV-tk transfected vascular endothelial cells, as well as to immune response-related phenomena (Vile et al. 1994).

1.5 Gene Transfer into APC or Bystander Cells

Because of the difficulty of culturing and transfecting autologous tumor cells, some investigators propose the use of bystander cells, e.g., fibroblasts, as the source of cytokine production in the generation of systemic antitumor immunity (Tahara et al. 1994; Veelken et al. 1994; Lotze et al. 1994; Fakhrai et al. 1995; Mertelsmann et al. 1995). Animal models suggest that "paracrine" secretion of cytokines by cytokine gene-transfected fibroblasts mixed with irradiated unmodified tumor cells can also induce antitumor immunity; however, in some models higher cytokine levels seem to be necessary.

Over the past decade, a series of T cell-recognized antigens have been described in tumor cells, among them developmental antigens such as the melanoma associated antigenes (MAGE) family, differentiation antigens, and structurally abnormal proteins such as mutated or overexpressed anti-/oncogenes (e.g., *ras*, p53) and breakpoint peptides (e.g.,

bcr/abl) (Van der Bruggen et al. 1991; Chen et al. 1992; Fenton et al.
1993). Carcinoembryonic antigen (CEA) is a well-characterized antigen
which, in the adult organism, is expressed by tumor cells and only to a
lesser extent by normal colonic epithelium (Thompson et al. 1991). In
mice, vaccination with this antigen generated a cellular immune response, which prompted the initiation of a clinical trial in patients with
colon cancer (Conry et al. 1994; Kantor et al. 1992). As tumor cells are
often defective in antigen processing or antigen presentation or even
actively secrete immunosuppressive factors, transfection of defined tumor antigens in professional APC such as dendritic cells offers the
possibity of circumventing this problem (Steinmann 1991). To even
increase the immune response stimulating function of APC, one could
cotransfect other molecules implicated in the generation or amplification of an immune response, such as cytokines or accessory molecules.
Preclinical experiments indicate that pulsing of APC with the appropriate antigenic peptide or coincubation with a tumor cell lysate can also
induce antigen-specific, MHC-restricted T cells (Inaba et al. 1990). The
latter approach in particular seems likely to facilitate clinical application
greatly since transfection would not be needed and even the necessicity
to know the sequence of the antigenic peptide could possibly be overcome. In vitro culture techniques that allow the generation of sufficient
numbers of highly purified human dendritic and Langerhans cells from
peripheral blood hematopietic stem cells will promote the development
of clinical protocols in the near future (Mackensen et al. 1995).

1.6 Chemo- and Radioprotection of Non-neoplastic Tissues

The protection of non-neoplastic cells is currently being pursued experimentally and clinically in patients with brain tumors, ovarian cancer,
and disseminated breast cancer by transferring the multidrug resistant
type 1 gene (MDR-1 gene) or other genes coding for detoxifying enzymes into hematopoietic progenitor cells. The MDR-1 gene codes for
an energy-dependent efflux pump for a great many cytotoxic anticancer
drugs and thus induces resistance towards these substances. In some
instances, dose escalation might be clinically indicated but cannot be
achieved because sensitive tissues such as the bone marrow limit cytotoxic therapy. By ex vivo transfection of the MDR-1 gene into he-

matopietic stem cells, these cells should, after transplantation and following high-dose myelosuppressive chemotherapy, be resistant to the toxic effects of subsequent chemotherapies (Sorrentino et al. 1992; Mickisch et al. 1992; Licht et al. 1995). One potential danger, however, could be that tumor cells contaminating the hematopoietic stem cell preparation would also be transfected with the MDR-1 gene and thus could also be rendered resistant against chemotherapy. Preclinical data suggest that this approach will be less toxic. In various preclinical models an increase in drug dose by 50% without damage to the hematopoietic system could be achieved. Furthermore, the ability to select for a drug resistance gene in vivo opens the possibility to cotransfect otherwise nonselectable genes into hematopoietic cells for the treatment of cancer and other diseases. Still under preclinical investigation are similar approaches in which antioxidants such as gluthatione-transferase or gluthatione-synthetase are overexpressed in non-neoplastic tissue to confer a radioprotective effect.

1.7 Systemic Production of Therapeutic Molecules by Genetically Modified Cells In Vivo

Another gene therapeutic strategy, not only for patients with cancer, is the systemic production of therapeutic molecules in vivo by genetically modified cells. Several diseases are characterized by a permanent or transient deficiency in certain biologically active proteins or other molecules which are required for important biological functions (Selden et al. 1987). For example, hemophiliacs lack an important clotting factor. Due to a usually short serum half-life the missing proteins often have to be replaced as a plasma product or in recombinant form by multiple daily or even continuous intravenous administration. This replacement therapy is not only tedious but also costly. An alternative approach for a permanent delivery of biologically active molecules in vivo would consist of injecting autologous or allogeneic cells which have been genetically modified to secrete the missing molecule. Transfection of suitable somatic cells with the relevant genes would be performed ex vivo or in vivo. Ex vivo transfection would allow clonal selection and then expansion of the transfected cells in vitro and a thorough characterization of the cell population reinjected into the patient. In a preclinical therapeutic murine model using GM-CSF-transduced fibroblasts, a single injection of irradiated GM-CSF-transduced fibroblasts was shown to be equally efficacious as twice daily s.c. injections for 7 days of the recombinant protein (Rosenthal et al. 1994c). A single injection of G-CS- secreting irradiated fibroblasts also led to accelerated hematologic recovery as well as mobilization of progenitor cells into the peripheral blood (Rosenthal et al. 1996). In many animal models, long-term expression of the transfected gene is not seen in vivo. On the one hand, this is attributed to a limited long-term cell survival as well as a shut-off of the transfected promoter in vivo. Clinically, the use of a so-called "immunoisolation device" in which genetically modified cells can be injected would have the advantage of allowing the use of allogeneic or even xenogeneic cells. Animal models have shown that cells inside the immunoisolation device are protected from destruction by a cell-mediated or antibody-mediated immune response. Also, at the appearence of any toxic side effects, the transfected cells can be removed by simply rinsing the chamber. Another safety step would be the cotransfection of a suicide gene, as described above, that would allow selective elimination of transfected cells in vivo after administration of the appropriate pro-drug.

1.8 Conclusions

The new developments in molecular biology in medicine as well as in cell biology and immunology have delineated new strategies for clinical cancer therapy. The specific genetic alterations in cancer cells are potential targets for drug, gene, or immunological therapies. At what point in time gene therapy would be clinically relevant in taking advantage therapeutically of the molecular differences between cancer cells and normal cells is difficult to predict. While many immunological and molecular strategies have already been elucidated in preclinical models, therapeutic efficacy can only be assessed in clincal trials. Therapeutic end points such as the regression of cancer lesions will be difficult to achieve in the early phases of clinical development of these new treatment strategies. In preliminary clinical trials, it will be important to define the biological end points, such as the demonstration of specific cytotoxic killer cells after therapeutic intervention. This will be extremely important for further rational developments of these promising beneficial approaches in treating patients with cancer. It is likely that the direction of clinical trials will progress into combining different therapeutic approaches in order to maximize therapeutic efficacy. Currently, gene therapeutic strategies are predominantly focussed on ex vivo manipulation of cultured cells. Future approaches are likely to focus on direct in vivo transfection of target cells, thus eliminating the requirement to isolate, transfect, culture, and reintroduce cells into the patient. For this, highly efficient and targeted gene delivery vectors must be developed.

References

Allione A, Consalvo M, Nanni P, Lollini PL, Cavallo F, Giovarelli M, Forni M, Gulino A, Colombo MP, Dellabona P, Hock H, Blankenstein T, Rosenthal FM, Gansbacher B, Bosco MC, Musso T, Gusella L, Forni G (1994) Immunizing and curative potential of replicating and nonreplicating murine mammary adenocarcinoma cells engineered with interleukin (IL)-2, IL-4, IL-6, IL-7, IL-10, tumor necrosis factor, granulocyte-macrophage colony-stimulating factor, and γ-interferon gene or admixed with conventional adjuvants. Cancer Res 54:6022–6026

Anderson WF (1984) Prospects for human gene therapy. Science 226:401-409

Blankenstein T, Qin Z, Überla K, Müller W, Rosen H, Volk HD, Diamantstein T (1991) Tumor suppression after tumor cell-targeted tumor necrosis factor α gene transfer. J Exp Med 173:1047–1052

Breakefield XO, DeLuca NA (1991) Herpes simplex virus for gene delivery to neurons. New Biol 3:203–218

Brenner MK, Rill DR, Moen RC, Krance RA, Mirro J, Anderson F (1993) Gene marking to trace origin of relapse after autologous bone marrow transplantation. Lancet 341:85–86

Chasse JF, Levrero M, Kamoun P, Minet M, Briand P, Perricaudet M (1989) Adenovirus as vector for gene therapy. Med Sci 5:331–337

Chen L, McGowan P, Ashe S, Johnston J, Li Y, Hellstrom I, Hellstrom KE (1994) Tumor immunogenicity determines the effect of B7 costimulation on T cell-mediated tumor immunity. J Exp Med 179:523–532

Chen W, Peace D, Rovira DK, You S-G, Cheever MA (1992) T-cell immunity to the joining region of p210[BCR-ABL] protein. Proc Natl Acad Sci USA 89:1468–1472

Connor J, Bannerji R, Saito S, Heston W, Fair W, Gilboa E (1993) Regression of bladder tumors in mice treated with interleukin 2 gene-modified tumor cells. J Exp Med 177:1127–1134

Conry RM, LoBuglio AF, Loechel F, Moore SE, Sumerel LA, Barlow DL, Curiel DT (1994) A carcinoembryonic antigen polynucleotide vaccine has in vivo antitumor activity. Gene Ther 1:1–7

Culver KW, Ram Z, Walbidge S, Ishi H, Olfield E, Blaese M (1992) In vivo transfer with retroviral vector producer cells for the management of brain tumors. Science 256:1550–1552

Dranoff G, Jaffe E, Lazenby A, Goulumbek P, Levitzky H, Brose K, Jackson V, Hamada H, Pardoll D, Mulligan RC (1993) Vaccination with irradiated tumor cells engineered to secrete granulocyte-macrophage colony-stimulating factor stimulates potent, specific, and long-lasting anti-tumor immunity. Proc Natl Acad Sci USA 90:3539–3543

Dunbar CE, Cottler-Fox M, O'Shaughnessy JA, Doren S, Carter C, Berenson R, Brown S, Moen RC, Greenblatt J, Stewart FM, Leitman SF, Wilson WH, Cowan K, Young NS, Nienhuis AW (1995) Retrovirally marked CD34-enriched peripheral blood and bone marrow cells contribute to long-term engraftment after autologous transplantation. Blood 11:3048–3057

Eglitits MA, Kantoff P, Gilboa E, Anderson WF (1985) Gene expression in mice after high efficiency retroviral-mediated gene transfer. Science 230:1395–1398

Fakhrai H, Shawler DL, Gjerset R, Naviaux RK, Koziol J, Royston I, Sobol RE (1995) Cytokine gene therapy with interleukin-2-transduced fibroblasts: effects of IL-2 dose on anti-tumor immunity. Hum Gene Ther 6:591–601

Felgner, PL, Gadek TR, Holm M, Roman R et al (1987) Lipofection: a highly efficient, lipid-mediated DNA-transfection procedure. Proc Natl Acad Sci USA 84:7413–7417

Fenton RG, Taub DD, Kwak LW, Smith MR, Longo DL (1993) Cytotoxic T-cell response and in vivo protection against tumor cells harboring activated ras proto-oncogenes. J Natl Cancer Inst 85:1294–1302

Flynn JL, Weiss WR, Norris KA, Seifert HS, Kumar S, So M (1990) Generation of a cytotoxic T-lymphocyte response using a Salmonella antigen-delivery system. Mol Microbiol 4:2111–2118

Fox BA, Culver KW, Cornetta K, Morgan RA, Spiess PF, Mule JJ, Kasid A, Blaese RM, Anderson WF, Rosenberg SA (1989) Retroviral gene transduction of murine tumor infiltrating lymphocytes: a new approach to study trafficking in vivo. FASEB J 3:3496

Freeman SM, Abboud CM, Whartenby KA, Packman CH, Koeplin DS, Moolten FL, Abraham GN (1993) The "bystander effect": tumor regression when a fraction of the tumor mass is genetically modified. Cancer Res 53:5274–5283

Friedman T (1992) A brief history of gene therapy. Nature Genet 2:93–98

Gale RP, Grosveld G, Canaani E, Goldman JM (1993) Chronic myelogenous leukemia: biology and therapy. Leukemia 7:653-658

Gansbacher B, Rosenthal FM, Zier K (1993) Retroviral vector-mediated cytokine-gene transfer into tumor cells. Cancer Invest 11:345-354

Gansbacher B, Bannerji R, Daniels B, Zier K, Cronin K, Gilboa E (1990a) Retroviral vector-mediated γ-interferon gene transfer into tumor cells generates potent and long lasting antitumor immunity. Cancer Res 50:7820–7825

Gansbacher B, Zier K, Daniels B, Cronin K, Bannerji R, Gilboa E (1990b) Interleukin 2 gene transfer into tumor cells abrogates tumorigenicity and induces protective immunity. J Exp Med 172:1217–1224

Gansbacher B, Houghton A, Gilboa E, Golde D, Livingston Ph, Minasian L, Rosenthal F, Oettgen H, Steffens T, Yang SY, Wong G (1992a) A pilot study of immunization with HLA-A2 matched allogeneic melanoma cells that secrete interleukin-2 in patients with metastatic melanoma. Hum Gene Ther 3:677–690

Gansbacher B, Motzer R, Houghton A, Gilboa E, Bander N, Golde D, Minasian L, Gastl G, Rosenthal F, Scheinfeld J, Yang SY, Wong G, Reuter V, Livingston P, Bosl G, Nanus D, Fair WR (1992b) A pilot study of immunization with interleukin-2 secreting allogeneic HLA-A2 matched renal cell carcinoma cells in patients with advanced renal cell carcinoma. Hum Gene Ther 3:691–703

Golumbek PT, Lazenby AJ, Levitzky HI, Jaffé LM, Karasuyama M, Baker M, Pardoll DM (1991) Treatment of established renal cancer engineered to secrete interleukin-4. Science 254:713–716

Gross G, Eshhar Z (1992) Endowing T cells with antibody specificity using chimeric T-cell receptors. FASEB J 6:3370–3378

Hock H, Dorsch M, Kunzendorf U, Qin Z, Diamantstein T, Blankenstein T (1993a) Mechanisms of rejection induced by tumor cell-targeted gene transfer of interleukin 2, interleukin 4, interleukin 7, tumor necrosis factor, or interferon. Proc Natl Acad Sci USA 90:2774–2778

Hock H, Dorsch M, Kunzendorf U, Überla K, Qin Z, Diamantstein T, Blankenstein T (1993b) Vaccinations with tumor cells genetically engineered to produce different cytokines: effectivity not superior to a classical adjuvant. Cancer Res 53:714–716

Hui KM, Sim T, Foo TT, Oei AA (1989) Tumor rejection mediated by transfection with allogeneic class I histocompatibility gene. J Immunol. 143:3835–3843

Hwu P, Yannelli J, Kriegler M, Anderson WF, Perez C, Chiang Y, Schwarz S, Cowherd R, Delgado C, Mule J (1993) Functional and molecular characterization of tumor-infiltrating lymphocytes transduced with tumor necrosis factor-alpha cDNA for the gene therapy of cancer in humans. J Immunol 150:4104–4115

Inaba K, Metlay JP, Crowley MT, Steinman RM (1990) Dendritic cells pulsed with protein antigens in vitro can prime antigen-specific, MHC-restricted T cells in situ. J Exp Med 172:631–640

Kantor J, Irvine K, Abrams S, Kaufman H, DiPietro J, Schlom J (1992) Antitumor activity and immune responses induced by a recombinant carcinoembryonic antigen-vaccinia virus vaccine. J Natl Cancer Inst 84:1084–1091

Kotin RM, Siniscalco M, Samulski RJ, Zhu X, Hunter L, Laughlin CA, McLaughlin S, Muzyczka N, Rocchi M, Berns KI (1990) Site-specific integration by adeno-associated virus. Proc Natl Acad Sci USA 87:2211-2215

Kozarsky KF, Wilson JM (1993) Gene therapy: adenovirous vectors. Curr Opin Genet Dev 3:499-503

Kuriyama S, Yoshikawa M, Ishizaka S, Tsujii T, Ikenaka K, Kagawa T, Morita N, Mikoshiba K (1991) A potential approach for gene therapy targeting hepatoma using a liver-specific promoter on a retroviral vector. Cell Struct Funct 16:503–510

Lange W, Cantin EM, Finke J, Dölken G (1993) In vitro and in vivo effects of synthetic ribozymes targeted against BCR/ABL mRNA. Leukemia 11:1786–1794

Licht T, Aksentijevich I, Gottesmann MM, Pastan I (1995) Efficient expression of functional human MDR1 gene in murine bone marrow after retroviral transduction of purified hematopoietic stem cells. Blood 86:111–121

Lotze MT, Rubin JT, Carty S, Edington H, Ferson P, Landreneau R, Pippin B, Posner M, Rosenfelder D, Watson C (1994) Gene therapy of cancer: a pilot study of IL-4-gene-modified fibroblasts admixed weth autologous tumor to elicit an immune response. Hum Gene Ther 5:41–55

Mackensen A, Herbst B, Kohler G, Wolff-Vorbeck G, Rosenthal FM, Veelken H, Kulmburg P, Schaefer HE, Mertelsmann R, Lindemann A (1995) Delineation of the dendritic cell lineage by generating large numbers of Birbeck granule-positive Langerhans cells from human peripheral blood progenitor cells in vitro. Blood 86:2699–2707

Maran A, Maitra RK, Kumar A, Dong B, Xiao W, Li G, Williams BRG, Torrence PF, Silverman RH (1994) Blockage of NF-kB signaling by selective ablation of an mRNA target by 2–5A antisense chimeras. Science 265:789–792

Mertelsmann R, Lindemann A, Boehm T, Brennscheidt U, Franke B, Kulmburg P, Lahn M, Mackensen A, Rosenthal FM, Veelken H (1995) Pilot study for the evaluation of T-cell mediated tumor immunotherapy by cytokine gene transfer in patients with malignant tumors. J Mol Med 34:1–2

Mickisch GH, Aksentijevich I, Schoenlein PV, Goldstein LJ, Galski H, Stahle C, Sachs DH, Pastan I, Gottesman MM (1992) Transplantation of bone marrow cells from transgenic mice expressing the human MDR1 gene results in long-term protection against the myelosuppressive effect of chemotherapy in mice. Blood 79:1087–1093

Miller DG, Adams MA, Miller AD (1990) Gene transfer by retrovirus vectors occurs only in cells that are actively replicating at time of infection. Mol Cell Biol 10:4239–4242

Moritz, D, Wels W, Mattern J, Groner B (1994) Cytotoxic T lymphocytes with a grafted recognition specificity for ERB B2-expressing tumor cells. Proc Natl Acad Sci USA 91:4318–4322

Moss B (1991) Vaccinia virus: a tool for research and vaccine development. Science 252:1662–1667

Mullen CA, Coale MM, Lowe R, Blaese RM (1994) Tumors expressing the cytosine deaminase suicide gene can be eliminated in vivo with 5-fluorocytosine and induce protective immunity to wild type tumor. Cancer Res 54:1503–1506

Nabel GJ, Nabel EG, Yang ZH, Fox BA, Plautz GE, Gao X, Huang L, Shu S, Gordon D, Chang AE (1993) Direct gene transfer with DNA-liposome complexes in melanoma: expression, biologic activity, and lack of toxicity in humans. Proc Natl Acad Sci USA 90:11307–11311

Pan ZK, Ikonomidis G, Lazenby A, Pardoll D, Paterson Y (1995) A recombinant Listeria monocytogenes vaccine expressing a model tumour antigen protects mice against lethal tumour cell challenge and causes regression of established tumours. Nature Med 5:471–477

Papadopoulos EB, Ladanyi M, Emanuel D, Mackinnon S, Boulad F, Carabasi MH, Castro-Malaspina H, Childs BH, Gillio AP, Small TN et al (1994) Infusions of donor leukocytes to treat Epstein-Barr virus-associated lymphoproliferative disorders after allogeneic bone marrow transplantation. New Engl J Med 330:1185–91

Pardoll D (1992) New strategies for active immunotherapy with genetically engineered tumor cells. Curr Opin Immunol 4:619-623

Proctor GN (1992) Microinjection of DNA into mammalian cells in culture: theory and practice. Methods Mol Cell Biol 3:209-231

Ram Z, Culver KW, Walbridge S, Blaese RM, Oldfield EH (1993) In situ retroviral-mediated gene transfer for the treatment of brain tumors in rats. Cancer Res 53:83–88

Riddell SR, Watanabe KS, Goodrich JM, Li CR, Agha ME, Greenberg PD (1992) Restoration of viral immunity in immunodeficient humans by the adoptive transfer of T cell clones. Science 257:238–241

Rill DR, Santans VM, Roberts WM, Nilson T, Bowman LC, Krance RA, Heslop HE, Moen RC, Ihle JN, Brenner MK (1994a) Direct demonstration that autologous bone marrow transplantation for solid tumors can return a multiplicity of tumorigenic cells. Blood 84:380–383

Rosenthal FM, Mertelsmann R (1997) Present strategies for gene therapy in cancer. Onkologie (in press)

Rosenthal FM, Zier KS, Gansbacher B (1994a) Human tumor vaccines: genetic engineering of tumors with cytokine and histocompatibility genes to enhance immunogenicity. Curr Opin Oncol 6:611–615

Rosenthal FM, Cronin K, Bannerji R, Golde DW, Gansbacher B (1994b) Augmentation of antitumor immunity by tumor cells transduced with a retroviral vector carrying the interleukin-2 and interferon-gamma cDNAs. Blood 83:1289–1298

Rosenthal FM, Früh R, Henschler R, Veelken H, Mackensen A, Kulmburg P, Gansbacher B, Mertelsmann R, Lindemann A (1994c) Cytokine therapy with gene transfected cells: single injection of irradiated granulocyte-macrophage colony-stimulating factor-transduced cells accelerates hematopoietic recovery after cyotoxic chemotherapy in mice. Blood 84:2960–2965

Rosenthal FM, Kulmburg P, Früh R, Pfeifer, C, Veelken H, Mackensen A, Köhler G, Lindemann A, Mertelsmann R (1996) Systemic hematologic effects of granulocyte colony-stimulating factor produced by irradiated gene-transfected fibroblasts. Hum Gene Ther 7:2147–2156

Saito S, Bannerji R, Gansbacher B, Rosenthal FM, Romanenko P, Heston WDW, Fair WR, Gilboa E (1994) Immunotherapy of bladder cancer with cytokine gene-modified tumor vaccines. Cancer Res 54:3516–3520

Selden RF, Skoskiewicz MJ, Howie KB, Russel PS, Goodman HW (1987) Implantation of genetically engineered fibroblasts into mice. Science 236:715–718

Shimada T, Fujii H, Mitsuya H, Nienhuis AW (1991) Targeted and highly efficient gene transfer into CD4$^+$ cells by a recombinant human immunodeficiency virus retroviral vector. J Clin Invest 88:1043–1047

Smith CA, NG CYC, Heslop HE, Holladay MS, Richardson S, Turner EV, Loftin SK, Li C, Brenner MC, Rooney CM (1995) Production of genetically modified Epstein-Barr virus-specific cytotoxic T cells for adoptive transfer to patients at high risk of EBV-associated lymphoproliferative disease. J Hematother 4:73–79

Snyder DS, Wu Y, Wang JL, Rossi JJ, Swiderski P, Kaplan B et al (1993) Ribozyme-mediated inhibition of bcr-abl gene expression in a Philadelphia chromosome-positive cell line. Blood 82:600-605

Sorrentino BP, Brandt SJ, Bodine D, Gottesman M, Pastan I, Cline A, Nienhuis AW (1992) Selection of drug-resistant bone marrow cells in vivo after retroviral transfer of human MDR1. Science 257:99–103

Stancovski I, Schindler DG, Waks T, Yarden Y, Sela M, Eshhar Z (1993) Targeting of T lymphocytes to Neu/HER2-expressing cells using chimeric single chain Fv receptors. J Immunol 151:6577–6582

Steinmann RM (1991) The dendritic cell system and its role in immunogenicity. Annu Rev Immunol:9 271–296

Sun WH, Burkholder JK, Sun J, Culp J, Turner J, Lu XG, Pugh TD, Ershler WB, Yang NS (1995) In vivo cytokine gene transfer by gene gun reduces tumor growth in mice. Med Sci 2889–2893

Tahara H, Zeh HJ, Storkus WJ, Pappo I, Watkins SC, Gubler U, Wolf SF, Robbins PD, Lotze MT (1994) Fibroblasts genetically engineered to secrete interleukin 12 can suppress tumor growth and induce antitumor immunity to a murine melanoma in vivo. Cancer Res 54:182–189

Tepper RI, Mule J (1994) Experimental and clinical studies of cytokine gene-modified tumor cells. Hum Gene Ther 5:153–164

Thompson J, Grunert F, Zimmerman W (1991) CEA gene family: molecular biology and clinical perspectives. J Clin Lab Anal 5:344–366

Torrence PF, Maitra RK, Lesiak K, Khamnei S, Zhou A, Silverman RH (1993) Targeting RNA for degradation with a (2'-5')oligoadenylate-antisense chimera. Proc Natl Acad Sci USA 90:1300-1304

Townsend SE, Allison JP (1993) Tumor rejection after direct costimulation of CD 8+ T cells by B7-transfected melanoma cells. Science 259:368–370

Van der Bruggen P, Traversari C, Chomez P, Lurquin C, de Plaen E, Van den Eynde B, Knuth A, Boon T (1991) A gene encoding an antigen recognized by cytolytic T lymphocytes on a human melanoma. Science 254:1643–1647

Veelken H, Jesuiter H, Mackensen A, Kulmburg P, Schultze J, Rosenthal FM, Mertelsmann R, Lindemann A (1994) Primary fibroblasts from human adults as target cells for ex vivo transfection and gene therapy. Hum Gene Ther 5:1205–1212

Vieweg J, Rosenthal FM, Bannerji R, Heston WDW, Fair WR, Gansbacher B, Gilboa E (1994) Immunotherapy of prostate cancer in the Dunning rat model: use of cytokine gene modified tumor vaccines. Cancer Res 54:1760–1765

Vile RG, Hart IR (1993a) In vitro and in vivo targeting of gene expression to melanoma cells. Cancer Res 53:962-967

Vile RG, Hart IR (1993b) Use of tissue-specific expression of the herpes simplex virus thymidine kinase gene to inhibit growth of established murine melanomas following direct intratumoral injection of DNA. Cancer Res 53:3860–3864

Vile RG, Nelson JA, Castleden S, Chong H, Hart IR (1994) Systemic gene therapy of murine melanoma using tissue specific expression of the HSVtk gene involves an immune component. Cancer Res 54:6228–6234

Yang G, Hellström KE, Hellström I, Chen L (1995) Antitumor immunity elicited by tumor cells transfected with B7-2, a second ligand for CD28/CTLA-4 costimulatory molecules. J Immunol:2794–2800

Yang NS, Burkholder J, Roberts B, Martinelli B, McCabe D (1990) In vivo and in vitro gene transfer to mammalian cells by particle bombardment. Proc Natl Acad Sci USA 87:9568–9572

Zenke M, Steinlein P, Wagner E, Cotten M, Beug H, Birnstiel ML (1990) Receptor-mediated endocytosis of transferrin-polycation conjugates: An efficient way to introduce DNA into hematopoietic cells. Proc Natl Acad Sci USA 87:3655–3659

2 Placental Blood:
New Options in Hematopoietic Stem Cell Transplantation

R.M. Weber-Nordt and R. Mertelsmann

2.1 Introduction

Placental tissue and neonatal cord have attracted the special attention of many different cultures throughout history, with some groups attributing deep religious importance to them. For example, in the nineteenth century, Indian tribes gave a gift of neonatal cord in turtle-shaped amulets to the newborns to ensure a long life.

It is only recently that researchers have discovered the importance of placental blood as a potential source of hematopoietic stem cells. Lethally irradiated mice could be rescued by the injection of neonatal mouse blood, demonstrating the existence of hematopoietic progenitors in cord blood (de La Sevelle et al. 1996).

E. Gluckman, in Paris, was the first to use human cord blood from a sibling donor to restore bone marrow functions in a 4 year old boy suffering from Fanconi's anemia (Gluckman et al. 1989). This revolutionary clinical result was the first to demonstrate the usefulness of placental blood in transplantation medicine. It took 3 more years of experience before cord blood could be used in a 6 year old boy with chronic myeloid leukemia to fully restore bone marrow functions after intensified regimens of high dose chemotherapy (Broxmeyer et al. 1992; Issaragrisil et al. 1992; Pahwa et al. 1994;Wagner et al. 1992). Since then cord blood has been used more than 300 times in patients suffering from inborn errors of hematopoiesis or immunological deficiencies such as Fanconi's anemia, aplastic anemia, thalassemia, sickle cell disease, X-linked lymphoproliferative syndrome and severe combined immunodeficiency disease (SCID) (Blaese and Culver 1992; Kohn et al. 1995; Kurtzberg et al. 1994; Wagner 1993). Furthermore, placental blood has been used as an alternative stem cell source for transplantation in patients with malignant diseases, e.g., acute lymphoid and myeloid leukemia, chronic myeloid leukemia and solid tumors such as neuroblastomas, after intensified regimens of chemotherapy (Broxmeyer et al. 1989; Kessinger 1992; Lu et al. 1996; Rubinstein et al. 1993). In children, cord blood has so far proven to be a realistic alternative stem cell source in addition to the clinically more established sources of autologous bone marrow or peripheral blood and allogeneic bone marrow (Broxmeyer et al. 1989; Hong and Deeg 1994; Lind 1994; Russell et al. 1993).

Although it is very early in the history of cord blood medicine, the advantages are manifold: (1) cord blood is readily accessible; (2) it is available to anyone; (3) can be obtained without any harm or pain to the mother or child after child and placental birth; and (4) contains higher numbers of hematopoietic stem cells than adult peripheral blood. Moreover the neonatal progenitor cells are readily expandable and highly proliferative in response to many growth factors and cytokines (Broxmeyer et al. 1992; Civin 1995). Thus, there are significant differences between cord blood and adult progenitor cell preparations with respect to a more advantageous source for hematopoietic stem cell separation (Mayani et al. 1993; Nimgaonkar et al. 1995).

The need for an alternative stem cell source in hematopoietic transplantation is demonstrated by the following: The use of autologous progenitor cells either from bone marrow or peripheral blood upon granulocyte or granulocyte/macrophage colony-stimulating factor (G-CSF/GM-CSF) treatment carries multiple risks such as (1) general anesthesia or growth factor treatment, (2) pain because of multiple bone punctures to obtain bone marrow as well as (3) the risk of transplanting back into the patient contaminating tumor cells, including those from solid tumors such as mammary carcinoma or neuroblastoma (Fraig 1992; Gribben and Nadler 1993). Even purging techniques (ribozymes, cellular chromatography) do not guarantee complete tumor-cell removal from the transplant, thereby leaving the risk of early or late relapses upon autologous stem-cell transplantation. However, autologous cells are available for most tumor patients as long as marrow functions are sufficient, and with autologous stem-cell transplantation, there is no risk of immunological failure of the transplant due to graft-versus-host disease.

By contrast, the use of allogeneic cells for stem cell transplantion does not carry the potential of malignant cell contamination (Matsunaga et al. 1993). However, while allogeneic cells of matched sibling donors are the most well-tolerated in the recipient, they are available for only 30% of patients. Some 40% of patients will find a matched unrelated donor, but the remaining one third of the patients do not find a suitable donor and cannot be transplanted within a reasonable amount of time. In addition, with the transplantation of unrelated cells – even with a fully matched HLA loci – there is a high chance of acute overall graft-versus-host disease leading to lethal outcome of the transplantation and to the death of the patient (Collins and Fernandez 1994). In a large multicenter study, Kernan et al. (1993) demonstrated the limitations of using unrelated allogeneic bone marrow in 492 patients with malignant or nonmalignant hematopoietic diseases. After a median time to transplantation of 196 days, $1–5 \times 10^8$ mononuclear cells were transplanted in a patient cohort (median age 22 years) resulting in an engraftment probability of initially 94%. Cell recovery was achieved within 22 days after transplantation. The overall disease-free survival was 40% for low risk patients, 19% for high risk patients and 37% for patients with chronic myeloid leukemia. The rate of relapse was 19%. Most interestingly, the overall death rate was 66% in 1.5 years with 33% of these deaths caused by severe graft-versus-host diseases. Other causes of death were infections (37%), toxicity (14%), graft failure (11%) or other (15%). In summary, the transplantation of hematopoietic stem cells currently requires a source of progenitor cells that: (1) is not contaminated with tumor cells, such as occurs with autologous cell preparations, and (2) does not induce severe graft-versus-host disease (GvHD)as do unrelated stem cells, and (3) is available in sufficient numbers so as to serve every patient. Recent clinical results indicate that neonatal blood derived from the placenta may fulfill these requirements (Rubinstein et al. 1993; Wagner et al. 1992; Wagner 1992). Clinical data from two independent studies, discussed below, support this idea.

2.2 Clinical Results

Wagner et al. (1995) presented evidence for the successful use of neonatal placental blood in the recovery of bone marrow functions in 44 children who were transplanted with their sibling-derived cord blood. Of the 44 patients 25 suffered from malignant and 19 from nonmalignant diseases. The median age was 4 years and the median weight 18.6 kg. In contrast to the much higher cell dose needed for adult stem cells, cord blood-derived stem cells became engrafted when a median of 5×10^7 (in a range of $1-33 \times 10^7$ cells) mononuclear cells per kilogram body weight were used. Patients were transplanted with fully matched grafts ($n=34$), with one mismatch accepted ($n=4$) or with two or three mismatches accepted in the HLA-A, B, C loci ($n=1$ and $n=5$, respectively). Within a median follow-up time of 1.6 years, comparable to the study presented by Kernan et al. (1993), an overall survival of 72% was observed. Causes of death were mainly infections, low frequency graft failures ($n=5$), GvHD ($n=1$) or acute respiratory distress syndrome (ARDS) ($n=2$). The overall prevalence of GvHD was 3% and was of low severity, responding well to corticoid treatment. Thus, this study demonstrated the feasibility of using cord blood in related patient recipients to restore bone marrow functions after high dose chemotherapy for the treatment of cancer.

A study by Kurtzberg et al. (1996) presented evidence for the clinical usefulness of cord blood in unrelated recipients. In 25 patients ($n=19$ malignant and $n=5$ nonmalignant hematologic diseases), with a median age of 7 years and a median body weight of 19.4 kg, a median cell dose of 3.6×10^7 mononuclear cells (range of $0.7-11 \times 10^7$ cells) was sufficient to provide engraftment and neutrophil recovery after a median of 22 days. Again, HLA mismatches were accepted in 24 of the 25 patients without increasing the risk of GvHD. The median time to find a donor was 115 days and the median follow-up time after transplantation was 12.5 months. Seven patients died because of infection ($n=4$), toxicity ($n=2$) or relapse ($n=1$). Most importantly, GvHD was not a cause of death.

In 24 of the 25 patients, no GvHD or low grade GvHD was observed in 19 patients, and five patients had grade III or IV GvHD that responded well to corticoid treatment and finally resolved. In summary, this study demonstrated a significantly lower frequency of severe grade GvHD and that GvHD did not influence the survival rate in patients transplanted with human umbilical cord blood. The relatively long time (115 days) needed to find a suitable graft can be explained by the low amount of cord blood that has been collected so far and underlines the importance of large cord blood banks all over the world to provide HLA-matched grafts for any ethnic minority and any HLA type.

A further report, by Laporte et al. (1996), described for the first time the use of cord blood in a 26 year old adult patient. Ten million mononuclear cells (MNCs) per kg resulted in the recovery of neutrophil counts above 500/ml after 23 days. Thus, bone marrow functions can be restored by numbers of MNC that are 10- to 40-fold lower than that of hematopoietic progenitor cells derived from adult bone marrow. The clinical data so far clearly demonstrate that umbilical cord blood-derived cells efficiently reconstitute bone marrow functions in patients with malignant and nonmalignant hematologic diseases. The use of cord blood-derived stem cells results in GvHD responses of low severity. GvHD in the setting of cord blood transplantation is not of life-threatening intensity but may provide enough reactivity to guarantee a graft versus leukemia (GvL) effect, which has been suggested to suppress relapse after transplantation. In the following, the characteristics of the main cell types responsible for the different results obtained with cord blood versus adult hematopoietic progenitor cell transplantation are discussed.

2.3 Hematopoietic Progenitor Cells

In vitro analyses of cord blood-derived hematopoietic progenitor cells provide evidence that CD34 antigen-expressing stem cells in cord blood contain an earlier fraction of progenitor cells that are rh123^{low}, CD38$^-$ and which have a high proliferative potential (Lu et al. 1993; Nimgaonkar et al. 1995; Tjonnfjord et al. 1994). When plated in methylcellulose colony assays, these cells form large colonies. Furthermore, these CD34+/CD38$^-$ cells provide long-term marrow reconstituting cells at high frequencies (Lu et al. 1993; Mayani et al. 1993; Migliaccio et al. 1992). As a side effect of their high proliferative potential and their elevated responsiveness to hematopoietic growth factors, cord blood-derived stem cells display a comparatively high efficiency of retroviral gene transfer (Blaese and Culver 1992; Hanley et al. 1994). Our own results clearly demonstrate that, by using the magnetic cell sorting system provided by Miltenyi Biotech, Bergish Gladbach, Germany), these CD34+ cells can be separated from cord blood to a purity of 99%. In methylcellulose colony-forming assays, these cells have a threefold higher response rate in the formation of myeloid colonies when treated with interleukin (IL)-3, G-CSF and erythropoietin (EPO) compared to bone marrow derived CD34+ cells. Of the colony-forming cells, 1 in 10^4 cells displays replating capability. With respect to long-term-culture-initiating cells (LTCICs), cord blood has been demonstrated to contain 2.5-fold higher levels than adult bone marrow.

Taking into consideration that neonatal blood carries hematopoietic stem cells from either the liver or the spleen to the bone marrow of the fetus, it becomes clear that, at the time of birth, neonatal blood contains the fetal cells that are on their way to populate the neonatal bone marrow. These early and immature hematopoietic progenitor cells very easily respond to proliferative signals provided by the addition of growth factors such as IL-3, IL-1, IL-6, G-CSF, stem cell factor (SCF), FLT3, and EPO. These cytokines were tested in vitro for their effects on cell proliferation and cell expansion. A combination of IL-3, SCF and FLT3 was the most potent in expanding cell numbers in the cell culture system used. Thus, in vitro results suggest that cell expansion techniques may be beneficial in increasing cell numbers and in retaining cell quality in the transplantation of adult recipients after myeloablative chemotherapy (Broxmeyer et al. 1992). Further studies are underway to determine the characteristics and cloning potentials of the expanded cord blood cells.

2.4 Immune Cells

Another striking difference between the in vivo use of adult hematopoietic progenitor cells, either from bone marrow or peripheral blood, and cord blood is the relative lack of severe forms of GvHD in patients transplanted with the latter (Garban et al. 1996; Risdon et al. 1994). GvHD is a combination of processes that involve diverse pathophysiological conditions (Przepiorka et al. 1995). Chemotherapy, which precedes transplantation of hematopoietic stem cells, induces tissue damage in the host and induces the release of proinflammatory cytokines, e.g., IL-1 and tumor necrosis factor-α (TNFα). Along with endotoxin, these cytokines are potent activators of host macrophages, that become activated to present antigens to surrounding T lymphocytes. These T lymphocytes, which are derived from the donor in a transplanted patient, are thus driven to further produce TH$_1$-like cytokines such as IL-12 or interferon (IFN)-γ, leading to acute graft-versus-host responses. Second, T lymphocytes are activated to produce TH$_2$-type cytokines, including IL-4 or IL-10, which primarily suppress acute GvHD but induce chronic GvHD. Cytokine dysregulation and the further activation of more potent antigen presenting cells in diverse tissues subsequently lead to targeted organ damage. Of the cytokines involved in this cascade IL-2 is important because of its ability to induce cytokine production in other immune cells and to recruit effector cells to the site of action (Przepiorka et al. 1995). Besides IL-2, it is (IFN)-γ which mediates direct cytopathic effects and which increases expression of MHC class I and II molecules on recruited cells of the immune system, thus leading to a mounted immune response. Due to this activity (IFN)-γ is thought to be the main effector of intestinal toxicity in the host (Przepiorka et al. 1995). This cytokine induces natural killer cell activity and the production of nitric oxid, further increasing cytopathic responses. By contrast, (IFN)-γ suppresses lymphocyte proliferation and thus leads to a state of immunodeficiency in the host, favoring the outbreak of opportunistic infections in the transplanted patient. Of the clinical symptoms that are associated with GvHD, IL-1 is mainly responsible for the induction of the wasting syndrome observed in the patients and TNF<a plays a significant role in the development of venoocclusive disease by inducing increased production of acute phase proteins. Thus, with regard to the reduced rate of GvHD reactions in patients transplanted with cord blood, one could hypothesize that immune cells in cord blood are less responsive to those factors that trigger GvHD. In vitro studies have demonstrated that indeed monocytes in cord blood display fewer MHC II molecules, resulting in a reduced capability to present antigen (Garban et al. 1996). Compared to adult peripheral blood-derived T lymphocytes, cord blood T cell functions are limited, i.e., helper cell activity, expression of surface molecules such as the IL-2 receptor, the production of IL-2 by activated T lymphocytes, and cyto-

toxic T lymphocyte activity (Harris et al. 1992; Roncarolo et al. 1994; Zola et al. 1995). Furthermore, cord blood-derived B lymphocytes produce low amounts of IgG or IgA as well as low numbers of MHC class II molecules (Garban et al. 1996). Thus, cord blood contains immature populations of immune cells which give rise to reduced immune responses in the presence of proinflammatory stimuli in the transplanted host.

2.5 Cord Blood Banking

Banking of cord blood may provide sufficient hematopoietic progenitor cells for allogeneic and/or autologous transplantation (Rubinstein et al. 1993). In addition to the clinical benefits of using cord blood summarized so far, a number of other benefits have to be mentioned. Cord blood is easy to obtain without any pain to the donor or risk of anesthesia or growth factor exposure. One single collection of cord blood contains sufficient cell numbers to be used in transplantation and the high proliferative quality of these stem cells indicates that even low cell numbers may be sufficient to ensure engraftment. Cord blood can be obtained in unlimited amounts and stored in liquid nitrogen without loss of cell viability for at least 8 years (Rubinstein et al. 1993). Compared to adult cells, cord blood contains a low level of infectious agents. Thus, it appears reasonable to collect and characterize cord blood for transplantation purposes. Cord blood banking comprises several steps: (1) cell collection, (2) cell processing, (3) cryopreservation, (4) analysis of cells and (5) data storage.

1. Cell collection: Cord blood can be collected within the third stage of labor with the placenta still in utero (Rubinstein et al. 1993). The alternative to cord blood collection is to obtain placental blood by gravity flow after the birth of both the child and the placenta (Rubinstein et al. 1993).
2. Cell processing: The final goal of further processing of the blood is to reduce volume and to lower storage costs. Volume reduction is sufficiently achieved by gravity sedimentation or centrifugation using hydroxyethylstarch followed by plasma extraction and separation of the white cells (Rubinstein et al. 1995). Cells are then further centrifuged and plasma is further separated (Rubinstein et al. 1995).

3. Cryopreservation: Processed cord blood cells can be cryopreserved in a final concentration of 10% DMSO, 2.5% human albumin and 5% dextran (Meagher and Herzig 1993; Rubinstein et al. 1995). Cells are initially frozen at –80°C and transferred thereafter to the liquid phase of a nitrogen tank.
4. Analyses: Cord blood cells are typed for HLA and blood group specificity. Microbiological as well as serological testing is performed for the titer of cytomegalovirus, Epstein-Barr virus, toxoplasmosis, syphilis, human T cell leukemia virus I, human immunodeficiency virus, herpes simplex virus, and hepatitis B and C viruses. Cell numbers and numbers of colony-forming cells are determined (Rubinstein et al. 1993, 1995).
5. Data storage: Data concerning blood analyses, the medical history of the parents as well as the informed consent of the parents are stored in a computerized data base according to federal safety recommendations for data storage.

In summary, based on clinical as well as in vitro data, transplantation of cord blood appears to be a new and alternative source of stem cells (Thomas 1995; Wagner 1992). Cord blood banking, once globally performed, should allow suitable progenitor cell grafts within a few days and lead to lower transplantation mortality because of significantly reduced rates of GvHD (Przepiorka et al. 1995). Thus, placental blood provides new perspectives for the transplantation of hematopoietic progenitor cells in humans.

References

Blaese RM, Culver KW (1992) Gene therapy for primary immunodeficiency disease. Immunodef.Rev. 3:329–349

Broxmeyer HE, Douglas GW, Hangoc G, Cooper S, Bard J, English D, Arny M, Thomas L, Boyse EA (1989) Human umbilical cord blood as a potential source of transplantable hematopoietic stem/progenitor cells. Proc Natl Acad Sci USA 86:3828–3832

Broxmeyer HE, Hangcoc G, Cooper S, Ribeiro RC, Graves V, Yoder M, Wagner J, Vadhan-Raj S, Benninger L, Rubinstein P, Broun ER (1992) Growth characteristics and expansion of human umbilical cord blood and estimation of its potential for transplantation in adults. Proc Natl Acad Sci USA 89:4109–4113

Civin CI, Small D (1995) Purification and expansion of human hematopoietic stem/progenitor cells.Ann NY Acad Sci 770:91–98

Collins NH, Fernandez JM (1994) T cell depletion and manipulation in allogeneic hematopoietic cell transplantation. Immunomethods 5:189–196

de La Selle V, Gluckman E, Bruley-Rosset M (1996) Newborn blood can engraft adult mice without inducing graft versus host disease across non H-2 antigens. Blood 87:3977–3983

Fraig M (1992) Purging in autologous and allogeneic bone marrow transplantation. Current Opin Oncol 4:264–271

Garban F, Ericson M, Roucard C, Rabian-Herzog C, Teisserenc H, Sauvanet E, Charron D, Mooney N (1996) Detection of empty HLA class II molecules on cord blood B cells. Blood 87:3970–3976

Gluckman E, Broxmeyer HE, Auerbach AD, Freidman HS, Douglas GW, Devergie A, Esperou H, Thierry D, Socie G, Lehn P, Cooper S, English D, Kurtzberg J, Bard J, Boyse EA (1989) Hematopoietic reconstitution in a patient with Fanconi anemia by means of umbilical cord blood from a HLA identical sibling. N Engl J Med 321:1174–1178

Gribben JG, Nadler LM (1993) Bone marrow purging for autologous bone marrow transplantation. Leuk Lymphoma 11:141–148

Hanley ME, Nolta JA, Parkman R, Kohn DB (1994) Umbilical cord blood cell transduction by retroviral vectors: preclinical studies to optimize gene transfer. Blood Cells 20:539–543

Harris DT, Schumacher MJ, Locascio J, Besencon FJ, Olson GB, DeLuca D, Shenker L, Bard J, Boyse EA (1992) Phenotypic and functional immaturity of human umbilical cord blood T-lymphocytes. Proc Natl Acad Sci 89:10006–10010

Hong DS, Deeg HJ (1994) Hematopoietic stem cells: sources and applications Med Oncol 11:63–68

Issaragrisil S, Visuthisakchai S, Suvatte V, Tanpaichitr VS, Chandanayingyong D, Schreiner T, Kanokpongsakdi S, Siritanaratkul N, Pjankijagum A (1992) Transplantation of cord blood stem cells into a patient with severe thalassemia. N Engl J Med 332:367–369

Kernan NA, Bartsch G, Ash RC, Beatty PG, Champlin R, Filipovich A, Grajewski J, Hansen JA, Henslee-Downey J, McCullough J, McGlave P, Perkins HA, Phillips G, Sanders J, Stroncek D, Thomas D, Blume KG (1993) Analysis of 462 transplantations from unrelated donors fascilitated by the National Marrow Donor Program. N Engl J Med 328:593–602

Kessinger A (1992) Peripheral blood derived stem cells can be successfully cryopreserved without using controlled-rate freezing. Prog Clin Biol Res 377:603-610

Kohn DB, Weinberg KI, Nolta JA, Heiss LN, Lenarsky C, Crooks GM, Hanley ME, Annett G, Brooks JS, el-Khoureily A (1995) Engraftment of gene modified umbilical cord blood cells in neonates with adenosine deaminase deficiency. Nature Medicine 1:1017–1023

Kurtzberg J, Graham M, Casey J, Olson J, Stevens CE, Rubinstein P (1994) Placental blood as a source of hematopoietic stem cells for transplantation into unrelated recipients. Blood Cells 20:275–284

Kurtzberg J, Laughan M, Graham ML, Smith C, Olson JF, Halperin EC, Ciocci G, Carrier C, Stevens CE, Rubinstein P (1996) The use of umbilical cord blood in mismatched related and unrelated hematopoietic stem cell transplantation. N Engl J Med 335:157–166

Laporte J-P, Gorin N-C, Rubinstein P, Lesage S, Portnoi M-F, Barbu V, Lopez M, DouayL, Najman A (1996) Cord blood transplantation from an unrelated donor in an adult with chronic myelogenous leukemia. N Engl J Med 335:167–190

Lind SE (1994) Ethical considerations related to the collection and distribution of cord blood stem cells for transplantation to reconstitute hematopoietic function.Transfusion 34:828–834

Lu L, Xiao M, Shen R-N, Grigsby S, Broxmeyer H (1993) Enrichment, characterization, and responsiveness of single primitive CD34 human umbilical cord bloodhematopoietic progentiors with high proliferative and replating potential. Blood 81:41–48

Lu L, Shen R-N, Broxmeyer HE (1996) Enrichment, characterization, and responsiveness of single primitive CD34 human umbilical cord blood hematopoietic progenitors with high proliferative and replating potential. Crit Rev Oncol Hematol 22:61–78

Matsunaga T, Sakamaki S, KohgoY, Ohi S, Hirayama Y, Niitsi Y (1993) Recombinant human granulocyte colony-stimulating factor can mobilize sufficient amounts of peripheral blood stem cells in healthy volunteers for allogeneic transplantation. Bone Marrow Transpl 11:103–108

Mayani H, Dragowski W, Landsdorp PM (1993) Characterization of functionally subpopulations of CD34 cord blood cells in serum-free long term cultures supplemented with hematopoietic cytokines. Blood 82:2664–2672

Meagher RC, Herzig RH (1993) Techniques of harvesting and cryopreservation of stem cells.Hematol Oncol Clin North Am 7:501–533

Migliaccio G, Migliaccio AR, Druzin ML, Giardina P-J, Zsebo KM, Adamson JW (1992) Long-term generation of colony-forming cells in liquid culture of CD34+ cord blood cells in the presence of recombinant human stem cell factor. Blood 79:2620–2627

Nimgaonkar MT, Roscoe RA, Persichetti J, Rybka WB, Winkelstein A, Ball ED (1995) A unique population of CD34+ cells in cord blood. Stem Cells 13:158–166

Pahwa RN, Fleischer A, Than S, Good RA (1994) Sucessful hematopoietic reconstitution with transplantation of erythrocyte-depleted allogeneic human umbilical cord blood cells in a child with leukemia. Proc Natl Acad Sci USA 91:4485–4488

Przepiorka D, Weisdorf D, Martin P, Klingemann H-G, Beatty P, Hows J, Thomas ED (1995) Prevention of graft-versus-host disease with anti-CD5 ricin A chain immunotoxin after CD3 depleted HLA-nonidentical marrow transplantation in pediatric leukemia patients. Bone Marrow Transplant 15:825–828

Risdon G, Gaddy J, Broxmeyer HE (1994) Allogeneic responses of human umbilical cord blood. Blood Cells 20:566–572

Rubinstein P, Rosenfield RE, Adamson JW, Stevens CE (1993) Stored placental blood for unrelated bone marrow reconstitution. Blood 81:1679–1690

Rubinstein P, Dobrila L, Rosenfield RE, Adamson JW, Migliaccio G, Migliaccio AR, Taylor P, Stevens CE (1995) Processing and cryopreservation of placental/umbilical cord blood for unrelated bone marrow reconstitution. Proc Natl Acad Sci USA 92:10119–10122

Russell NH, Hunter A, Rogers S, Hanley J, Anderson D (1993) Peripheral blood stem cells as an alternative to marrow for allogeneic transplantation. Lancet 341:1482

Roncarolo M-G, Bigler M, Ciuti E, Martino S, Tovo P-A (1994) Immune responses by cord blood cells. Blood Cells 20:573–586

Thomas ED (1995) History current results, and research in marrow transplantation. Perspect Biol Med 38:230–237

Tjonnfjord GE, Steen R, Evensen SA, Thorsby E, Egeland T (1994) Evidence for engraftment of donor-type multipotent CD34+ cells in a patient with selective T-lymphocyte reconstitution after bone marrow transplantation for B-SCID. Blood 84:2795–27801

Wagner JE, Kernan NA, Steinbruch M, Broxmeyer HE, Gluckman E (1995) Allogeneic sibling umbilical cord blood transplantation in children with malignant and non-malignant disease. Lancet 346:214–219

Wagner J, Broxmeyer HE, Byrd RL, Zehnbeuer B, Schmekpeper B, Shah N, Griffin C, Emanuel PD, Zuckerman KS, Cooper S, Carow C, Bias W, Santos GW (1992) Transplantation of umbilical cord blood after myeloablative therapy: analysis for engraftment. Blood 79:1874–1881

Wagner JE (1992) Umbilical cord and placental blood hematopoietic stem cells: collection, cryopreservation, and storage. J Hematother 2:225–228

Wagner JE (1993) Umbilical cord blood stem cell transplantation. Am J Ped Hem Oncol 15:169–174

Zola H, Fusco M, Macardle PJ, Flego L, Roberton D (1995) Expression of cytokine receptors by human cord blood lymphocytes: comparison with adult blood lymphocytes. Ped Res 38:397–403

3 Recent Developments in the Ex Vivo Manipulation of Hematopoietic Cells from Bone Marrow and Blood

R. Henschler, J. Winkler, D. Möbest, A. Spyridonidis, W. Lange, and R. Mertelsmann

3.1 Introduction

Transplantation of the hematopoietic system has become a more and more widely used tool within antitumor treatment regimens. In leukemias, the primary goal is to completely eradicate the malignant clone; however, autologous hematopoietic stem cell transplantation has also been introduced as treatment for a number of solid tumors in an effort to increase chemotherapy intensity beyond levels that are conventionally dose limiting due to life-threatening bone marrow toxicity. Recombinant hematopoietic growth factors have been of substantial benefit in improving engraftment kinetics (Lieschke 1992, for review). Especially the mobilization of progenitor cells from bone marrow into the peripheral blood has triggered a substantial extension of application fields for hematopoietic stem cell transfer, which now includes a variety of solid tumor disease states. This was mainly due to the much simplified logistics of harvesting peripheral blood progenitor cells (PBPC) without the need of general anesthesia . In most instances, the peripheral blood turned out to be a superior source of transplantable stem cells compared to progenitor cells from bone marrow, especially as the speed of neutrophil and platelet recovery was hastened (Brugger et al. 1993b, 1994b, 1995; Elias et al. 1992; Sheridan et al. 1992; To et al. 1992). A number of studies are now underway investigating the clinical benefit for patients transplanted with PBPC in a variety of new dose-intensive chemotherapy protocols in different disease states, including non-Hodgkin's lymphomas, Hodgkin's disease, acute leukemias, and also some chemosensitive solid tumors, and some reports have already indicated improvements in antitumor cell response or survival (Bezwoda et al. 1995; Kessinger and Armitage 1991; Peters 1995).

In view of the increased effiency of in vivo tumor eradication by dose-intensive chemotherapy supported by stem cell transplantation, increasing concern has been raised about the fact that bone marrow or PBPC might still contain tumor cells in a proportion of patients with certain solid tumors or lymphomas (Brugger et al. 1994; Ross et al. 1993; Sharp et al. 1991, 1992). Since the marrow-repopulating stem cells reside in a subfraction that carries the CD34 surface antigen and comprises only about 1%–3% of bone marrow cells or PBPC, enrichment for progenitor cells by the surface marker CD34 has been introduced to deplete PBPC samples of tumor cells (Berenson et al. 1988). The development of high-affinity CD34 antibodies has led to the design of immunoaffinity systems that are now widely used in clinical trials (Berenson et al. 1991; Brugger et al. 1994).

3.2 Stem Cell Purification and Ex Vivo Expansion

In addition to the selection of stem cells by surface antigen-mediated antibody affinity purification techniques, an alternative or additional means of tumor cell depletion should be feasible if hematopoietic cells are maintained in liquid culture in the presence of hematopoietic growth factors. The principle is that the in vitro environment created selectively favors the expansion of hematopoietic cells, while unwanted malignant cells of nonhematopoietic origin are depleted by cell death due to adverse survival conditions. Before devices for enriching primitive hematopoietic cell populations on a large scale became available, growing cell fractions derived from bone marrow was notoriously difficult. The main reasons for this may have been the relatively low stem cell content of unseparated mononuclear cell populations and the ability of mature cells present in these populations to produce a number of cytokines negatively interfering with progenitor growth. The development of clinical grade CD34+ selection generating highly enriched CD34+ populations has been a major step forward in improving subsequent ex vivo expansion of hematopoietic cells in liquid culture. Another difficulty was that fewer different growth factors had been cloned at the time and thus were available as recombinant material that could be used to drive hematopoietic cell proliferation in culture. In addition, the advent of stem cell factor initiated a major change of this situation in 1990 (Witte 1990).

A second rationale for ex vivo expansion of hematopoietic cells is to provide amplified numbers of progenitor cells in a transplant, assuming that committed progenitors are mediators of early PBPC engraftment (Table 1). This may be of importance to further shorten neutropenic periods in patients during postchemotherapy aplasia. Progenitor ampli-

Table 1. Rationales for ex vivo expansion of hematopoietic progenitor cells

– To deplete of tumor cells contaminating the PBPC preparation
– To amplify numbers of colony forming cells to enforce engraftment
– To amplify intermediate stages of hematopoietic differentiation
  underrepresented in peripheral blood progenitor cell (PBPC) autografts
– To generate sufficient numbers of CFC from patients with low PBPC yields
– To avoid leukapheresis or bone marrow harvest under general anaesthesia

fication may also be of benefit when stem cell yields in PBPC or bone marrow harvests are low if sufficient numbers of progenitor cells can be generated by ex vivo expansion. In this case, ex vivo expansion may start from relatively small amounts of blood as starting material (500 ml or less), obviating the need for leukapheresis or bone marrow harvest procedures.

3.3 Protocols for Ex Vivo Expansion Cultures

Ex vivo culture of blood forming stem cells was first successfully achieved by Dexter and coworkers in the 1970s, when they discovered that the presence of a stromal layer, hydrocortisone, preselected batches of horse serum, and a culture temperature of 33°C were critical for the maintenance of transplantable stem cells and continuous development of functional mature cells (Dexter et al. 1977). A major development that formed the basis for a stroma-free, cytokine-supported method of ex vivo expansion of human bone marrow, peripheral blood, or cord blood stem cells was the description of a stroma-free, continuous culture system by Moore and coworkers (Moore 1991). This system, the "Delta Assay," grows progenitor cells over periods of several weeks in the presence of hematopoietic cytokines and can be used to assay the progenitor content of stem cell grafts. Depending on the quality and quantity of growth factors added, different plateau levels in the maximum achieved number of colony-forming cells (CFC) are generated in these cultures from a given stem cell population. To a high degree, these cultures reflect the different proliferation capacities of early, uncommitted, or partially committed progenitors in response to colony-stimulating factors (CSFs). For example, stem cell factor (SCF) acts mainly as a survival factor when present alone or increases proliferation and differentiation of stem cells in synergy with other CSFs; interleukin (IL)-1β acts mainly on progenitors with high proliferative capacity (Moore 1991). By differentially stimulating subsets of hemtopoietic progenitors, and by creating a spectrum of progenitor cell responses toward cell differentiation, the numbers of colony-forming progenitor cells may be amplified by up to about 2 logs over a period of 2–3 weeks. By regulating the balance between definitive induction of progenitors to differentiate and cell division that allowed, at least to some limited degree, self-renewal of primitive cells, these studies formed the methodological basis to further develop stroma-free, cytokine-supported hematopoietic cultures for use in therapy.

Generally, these cultures contained fetal calf serum. Haylock et al. (1992) used purified CD34+ PBPC cells as starting material and a combination of up to six hematopoietic growth factors. They readily obtained a more than 2-log amplification of total cell numbers over a 2-week culture period, consisting predominantly of immature neutrophil-like cells; no expansion of CFC was noted. These authors focused their approach primarily on the generation of neutrophil precursors in a state of maturation more terminal than the CFC, as this maturational state of precursors is not normally present in substantial numbers in PBPC harvests and may thus be able to confer production of mature neutrophils within a shorter interval posttransplantation than seen with PBPC. In a liquid culture system of PBPC CD34+ cells established at our institution, amplification of CFC by a factor of up to 190-fold was measured after 2 weeks of culture in a combination of five growth factors, SCF, IL-1, IL-3, IL-6, and erythropoietin (EPO) in the presence of fetal calf serum (Brugger et al. 1993a). In this system, the main emphasis was to amplify the number of CFC, since a number of studies transplanting PBPC had described inverse correlations between the number of CFC reinfused into patients and the period of absolute neutropenia (Pettengell et al. 1992).

3.4 Starting Cell Populations

From the studies mentioned above, it became clear that cultures derived from CD34+ cells showed a much more sustained cell proliferation of up to 6 weeks in culture than cultures initiated with unseparated mononuclear cell fractions, provided the cultures were refed and cells diluted at, in most cases, weekly intervals. Attempts have therefore been made to identify by more extensive enrichment techniques using surface antigen markers, such as flow cytometric cell sorting, more suitable populations of primitive progenitor cells. For example, CD34+, HLA-DR-, CD15- Rhodamin^{dull} cell populations separated by flow cytometry were able to further enhance the maximum increase in cell numbers in expansion cultures to more than 1500-fold – expanding at the same time different types of progenitor cells, including high proliferative potential-CFC (Srour et al. 1993). Among three different sorted stem cell populations from human umbilical cord blood, one was found to contain predominantly multipotent progenitors (CD34+ CD45RAlow CD71low), one to contain predominantly myeloid progenitors (CD34+ CD45RA+ CD71low), and one to contain predominantly erythroid progenitors (CD34+ CD45RAlow CD71+). The population containing most of the multipotent precursors (CD34+ CD45RAlow CD71low) showed the highest expansion rate of CD34+ cell numbers after 2 weeks of culture in a growth factor-supplemented serum-free medium (532-fold versus 130- and 28-fold, respectively (Mayani et al. 1993a). Another surface molecule which is potentially useful to further enrich for primitive cells from this cell population is the Thy-1 antigen (Baum et al. 1992).

During their ontogeny, hematopoietic stem cells display a decrease in their in vitro proliferative potential (Lansdorp et al. 1993). Concordant with this observation, cord blood stem cells have been shown to be a potentially very attractive starting cell population for ex vivo expansion (Broxmeyer et al. 1992; Cicuttini et al. 1992; Mayani et al. 1993a,b). However, for therapeutic use of expanded cells in autologous transplantation, PBPC (or bone marrow) may be the source of highest relevance because cord blood is usually not available for autologous transplantation of patients.

3.5 Stem Cell Assays To Control for the Presence of Stem Cells After Ex Vivo Expansion

Long-term bone marrow culture has been considered the most useful biologcal in vitro assay available, measuring cells potentially capable of long-term hematopoietic reconstitution (Testa and Dexter 1991). In mice, long-term bone marrow culture-initiating cells are almost congruent in terms of incidence and surface marker characteristics to bone marrow repopulant stem cells (Ploemacher et al. 1991). In our ex vivo expanded CD34+ PBPC, we have analyzed the content of "long-term bone marrow culture-initiating cells" (LTCIC) (Henschler et al. 1994). Overall, LTCIC could not be amplified during ex vivo expansion. They could, however, be maintained approximately at input levels if CSF,

IL-1, IL-3, IL-6 and EPO were all present during the expansion culture. Another method, fluorescent dye tracking, identified a fraction of primitive human bone marrow stem cells that were not recruited into cell division by a combination of IL-3, IL-6 and SCF, and EPO in culture (Lansdorp and Dragowska 1993). At the same time, ex vivo expansion took place in these cultures from more committed progenitor cells. Verfaillie et al. (1994) identified a culture medium composed of IL-3, macrophage-inflammatory protein 1 , and unknown activities secreted from stromal cells that was able to maintain LTCIC in culture for at least 8 weeks.

3.6 Hematopoietic Recovery in Therapeutic Application of Ex Vivo Expanded Cells

Studies using stroma-free, cytokine-supported ex vivo expansion protocols in mice found that an engrafting defect was induced upon liquid culture of hematopoietic progenitors (Peters et al. 1995, 1996). Before 1995, autologous stem cell grafting was only reported in humans after culture of bone marrow under stroma-forming culture conditions (Barnett et al. 1993; Chang et al. 1989), where a sometimes slow, but in all cases eventually successful hematopoietic recovery was obtained. At our institution, stroma-free, cytokine-expanded CD34+ PBPC were re-infused into patients following marrow-suppressive dose escalated chemotherapy in a phase I study (Brugger et al. 1995). In the presence of SCF, IL-1, IL-3, IL-6, EPO, and autologous serum 1/10 of a single 2-h leukapheresis product was expanded for 2 weeks. The reconstitution pattern in the patients receiving more than 1×10^5 CFU-GM/kg was identical to patients who had previously receiving CD34+ selected or unselected PBPC. Periods of absolute neutropenia and thrombopenia were between 4 and 7 days and thus were significantly (approximately 1 week) shorter than in controls from past studies who had not received stem cell support, but had had identical high-dose chemotherapy and growth factor support postchemotherapy (Table 2). This indicates that ex vivo expanded cells may be useful to confer (at least) short-term engraftment after the high-dose chemotherapy protocol mentioned above which normally results in an aplasia period of approximately

Table 2. Recovery times using PBPC and ex vivo expanded cells post high-dose chemotherapy with VP-16, ifosfamide, and cisplatin in patients with solid tumors

Source of transplant	CSFs post chemotherapy	Neutrophils <100/µl (days)	Platelets <20000/µl (days)	Patients (*n*)
–	G-CSF	10.5	8	6[a]
PBPC	G-CSF	6.5	3	8[a]
Ev vivo expanded CD34+ PBPC	G-CSF	7	4	4[b]

Data from (Brugger et al. 1993b, 1994b, 1995). All values represent median values.

[a]Patients received high dose chemotherapy with VP-16 (1500 mg/m^2), ifosfamide (12 g/m^2), and cisplatin (150 mg/m^2).

[b]Patients received high-dose chemotherapy with VP-16 (1500 mg/m^2), ifosfamide (12 g/m^2), carboplatin (750 mg/m^2), and epirubicin (150 mg/m^2).

3 weeks without progenitor cell reinfusion and growth factor application.

Since the chemotherapy protocol used in this patient group was not myeloablative, it is not known to which degree ex vivo expanded blood stem cells may confer long-term engraftment. These studies have not been performed in humans yet. Muench et al. (1993) transplanted mice with enriched bone marrow stem cells expanded ex vivo by IL-1, IL-6, and SCF. They demonstrated long-term donor-derived hematopoiesis in the transplanted animals. Similarly, Rebel et al. (1994) have observed maintenance and amplification of primitive hematopoietic stem cells defined by the expression of surface markers Sca-1 and wheat germ agglutinin, which were negative for lineage markers, in a serum-free culture system from murine bone marrow, starting from an enriched population of stem cells.

3.7 Systematic Improvement of Culture Conditions in Bioreactor Systems

Besides the capability of recombinant hematopoietic growth factors combinations to optimally support growth and development in ex vivo expansion cultures, increasing attention has been being paid to technical improvements in culture conditions. This includes the effects of continuous medium flow, rapid medium exchange, ion, lipid and protein content in the cultures, oxygen tension, and other components in the liquid phase of the culture system, as well as growth of progenitor cells on specially designed surface structures. Compared to serum-containing cultures, the use of serum-free medium resulted in more reliable and overall enhanced cell production of up to 400-fold within 12 days in our studies. A forced medium exchange (50% medium change every day) resulted in 2.5-fold better amplification rates in a system used by Schwartz et al. (1991). This correlated with an increased glucose consumption in these cultures (Caldwell et al. 1991). Koller et al. noted that the presence of a stroma layer and increased feeding schedules had a more beneficial effect on amplification or survival of CFU-GM and LTCIC than did alterations of cytokine concentrations or the cytokine combinations (Koller et al. 1995).

A logical next step from the results with enhanced feeding protocols was the use of perfused culture systems. Zandstra et al. (1994) achieved substantially higher cell yields, as compared to static cultures, simply by adding a continuously rotating stirrer to their culture bottle. In a system used by Koller et al. (1993a) it was noted that the depletion of glucose and IL-3, as well as the accumulation of lactate and IL-6 observed in static cultures, could be circumvented in continuously perfused cultures. These authors also found that an oxygen content of 5% was superior to the normal oxygen content of 20% to achieve maximum progenitor and total cell expansion (Koller et al. 1993). Moreover, continuous perfusion cultures seemed to enhance an expansion of LTCIC too.

The three-dimensional structure of the culture surfaces has been another focus of research in this field. Oh et al. (1994) saw extended growth when newly produced cells were harvested periodically from the bioreactor and assumed that enhancement of the available growth surface was responsible for this effect. A three-dimensional artificial microenvironment was introduced by Naughton et al. (1991) using suspended nylon screens that could sustain bone marrow cultivation for 270 days in a model using rat marrow cells and for 12 weeks using human bone marrow cells. Its usefulness for ex vivo expansion has not been analyzed yet, however. Sardonini and Wu (1993) described a hollow fiber system consisting of porcine microvascular endothelial cells, improving ex vivo expansion of human bone marrow CD34+ cells. Similarly, the use of two chambers separating the hematopoietic cells from continuously flowing medium by dialysis membranes created a microenvironment allowing continuous proliferation and maturation of, in this case, murine bone marrow cells (Wang et al. 1995). Again, this system has not been tested for ex vivo expansion. One may speculate that in the future, these aspects of culture technique will gain increasing importance. Another trend will probably be the development of handier, technically easy and uncomplicated systems, uniting the most important characteristics of the bioreactors mentioned above.

3.8 Advanced Technologies in Tumor Cell Purging Techniques

Carcinoma cells have been demonstrated within autolgous bone marrow but also in PBPC harvests (Brugger et al. 1994; Ross et al. 1993; Sharp et al. 1991, 1992). We assessed whether tumor cells could be depleted in our ex vivo expansion cultures and found a depletion of about tenfold, provided that serum-free medium was used (unpublished results). Also, the addition of transforming growth factor (TGF) β-1, an inhibitor of proliferation in a wide variety of cell types derived from many different tissues, resulted in increased cell death of mammary carcinoma cells and in their differentiation, while in the presence of the hematopoietic cytokines IL-1, IL-3, IL-6, SCF and EPO, hematopoietic progenitor cell expansion was still possible, although to a lesser degree. However, the absolute numbers of primitive (LTCIC) cells were not reduced in TGF-β-1 treated ex vivo expansion cultures compared to control cultures. Additional technology available for tumor cell purging includes the use of immunotoxins, i.e., monoclonal antibodies against tumor- or epithelial-specific antigens coupled to potent cellular toxins, and intracellu-

Table 3. Tumor cell purging strategies during ex vivo expansion of hematopoietic grafts

Method	Mechanisms of tumor cell depletion
Addition of hematopoietic cytokines; use of serum free medium	Elimination via selective tumor cell survival/growth disadvantage
addition of TGF-β	Tumor cell differentiation
Antisense ribozymes	Targetting of leukemia cell-specific oncogenic fusion transcripts (bcr-abl in chronic myelogenous leukemia)
Immunotoxins	Specific elimination of tumor cells by fusion molecule between antibody against epithelial cell specific surface markers (e.g., EGF receptor) and cell toxin (e.g., ricin A; *Pseudomonas* toxin)

TGF, transforming growth factor; EGF, epidermal growth factor.

lar molecular targeting using, for example, ribosomes against the bcr-abl fusion protein – which is of crucial importance in the proliferation of chronic myeloid leukemia (CML) progenitor cells and might allow the suppression of CML hematopoiesis in ex vivo culture of CML progenitors (Table 3).

3.9 Gene Transfer into Hematopoietic Stem Cells

The transduction of foreign DNA sequences into hematopoietic stem cell is a major therapeutic goal in the correction of inborn errors of metabolism, or the introduction of genes that confer resistance to cytotoxic drugs, which may allow further dose escalation of chemotherapy after successful bone marrow transplantation and stem cell gene transfer (Gieselmann 1995; Maze et al. 1996). Engraftment of transduced human cells into SCID mice was taken as an endpoint. The introduction of marker DNA into human stem cells has been successfully achieved with various retroviral vectors and transductions procedures using stroma-supported or stroma-free culture of hematopoietic cells, and the presence of hematopoietic cytokines has found to be of importance in the case of bone marrow cells, but not PBPC (Nolta et al. 1995). In monkeys or humans, the situation seems more difficult, and reports are still sparse which confirm the ongoing expression of therapeutic genes in excess of levels around 1%. It is not clear whether the defect lies in the insufficient transduction of the repopulating stem cells or whether the efficiently transduced stem cells do not home to the bone marrow or are eliminated by the immune system due to their genetic alteration, e.g., via presentation of antigen related to the newly expressed foreign proteins.

3.10 Perspectives

Hematopoietic progenitor cell expansion has been pursued as a new form of cellular therapy mainly for three reasons: Firstly, to increase the quality of stem cell grafts; secondly, to deplete autologous grafts from tumor cells; and thirdly, to allow efficient transduction of therapeutic genes into long-term bone marrow-repopulating stem cells. Tumor cell depletion has been achieved to a high degree in solid tumors, and this application will probably receive more attention in therapy with the introduction of specific molecular targeting technology, such as immunotoxins and antisense ribozyme technology. The conditions for maintaining long-term engrafting stem cells in culture need to be more fully characterized. Also, the actual feasibility of gene marking protocols and therapeutic gene transfer into hematopoietic stem cells in humans has not yet been completely evaluated; this includes applications in a context of improved cancer chemotherapy.

References

Barnett MJ, Eaves CJ, Phillips GL, Gascoyne RD, Hogge DE, Horsman DE, Humphries RK, Klingemann HG, Lansdorp PM, Nantel SH et al (1993) Autografting with cultured marrow in chronic myeloid leukemia: results of a pilot study. Blood 84:724–732

Baum CM, Weissman IL, Tsukamoto AS, Buckle AM, Peault B (1992) Isolation of a candidate human hematopoietic stem-cell population. Proc Natl Acad Sci USA 89:2804–2808

Berenson RJ, Andrews RG, Bensinger WI, Kalamasz D, Knitter G, Buckner CD, Bernstein ID (1988) Antigen CD34+ marrow cells engraft lethally irradiated baboons. J Clin Invest 81:951–955

Berenson RJ, Bensinger WI, Hill RS, Andrews RG, Garcia-Lopez J, Kalamasz D, Still BJ, Spitzer G, Buckner CD, Bernstein ID, Thomas ED (1991) Engraftment after infusion of CD34+ marrow cells in patients with breast cancer or neuroblastoma. Blood 77:1717–1722

Bezwoda WR, Seymour I, Dansey RD (1995) High dose chemotherapy with hematopoietic rescue as primary treatment for metastatic breast cancer: a randomized trial. J Clin Oncol 13:2483–2489

Broxmeyer HE, Hangoc C, Cooper S, Ribeiro RC, Graves V, Yoder M, Wagner J, Vadhan-Raj S, Benninger L, Rubinstein P et al (1992) Growth characteristics and expansion of human umbilical cord blood and estimation of its potential for transplantation in adults. Proc Natl Acad Sci USA 89:4109–4113

Brugger W, Mocklin W, Heimfeld S, Berenson RJ, Mertelsmann R, Kanz L (1993a) Ex vivo expansion of peripheral blood CD34+ progenitor cells by stem cell factor, interleukin-1 beta (IL-1 beta), IL-6, IL-3, interferon-gamma, and erythropoietin. Blood 81:2579–2584

Brugger W, Birken R, Bertz H, Frisch J, Schulz G, Mertelsmann R, Kanz L (1993b) Peripheral blood progenitor cells mobilized by chemotherapy + G-CSF accelerate both neutrophil and platelet recovery after high dose VP16, ifosfamide and cisplatin. Br J Haematol 84:402

Brugger W, Bross KJ, Glatt M, Weber F, Mertelsmann R, Kanz L (1994a) Mobilization of tumor cells and hematopoietic progenitor cells into peripheral blood of patients with solid tumors. Blood 83:636

Brugger W, Henschler R, Heimfeld S, Berenson R, Mertelsmann R, Kanz L (1994b) Positively selected autologous blood CD34+ cells and unseparated peripheral blood progenitor cells mediate identical hematopoietic engraftment after high-dose VP-16, ifosfamide, carboplatin, and epirubicin. Blood 84:1421–1426

Brugger W, Heimfeld S, Berenson RJ, Färber L, Mertelsmann R, Kanz L (1995) Reconstitution of hematopoiesis after high-dose chemotherapy by autologous progenitor cells expanded ex vivo. N Engl J Med 333:283–287

Caldwell J, Palsson BO, Locey M, Emerson SG (1991) Culture perfusion schedules influence the metabolic activity and granulocyte-macrophage colony stimulating factor production rates of human marrow stromal cells. J Cell Physiol 147:344–357

Chang J, Morgenstern GR, Coutinho LH, Scarffe JH, Carr T, Deakin DP, Testa NG, Dexter TM (1989) The use of bone marrow cells grown in long-term culture for autologous bone marrow transplantation in acute myeloid leukemia: an update. Bone Marrow Transplant 4:5–9

Cicuttini FM, Martin M, Salvaris E, Ashman L, Begley CG, Novotny J, Maher D, Boyd AW (1992) Support of human cord blood progenitor cells on human stromal cell lines transformed by SV40 large T antigen under the influence of an inducible (metallothionein) promoter. Blood 80:102–112

Dexter TM, Allen TD, Lajtha LG (1977) Conditions controlling the proliferation of hematopoietic cells in vitro. J Cell Physiol 91:335–344

Elias AD, Ayash L, Anderson KC et al (1992) Mobilization of peripheral blood progenitor cells by chemotherapy and granulocyte-macrophage colony-stimulating factor for hematologic support after high-dose intensification for breast cancer. Blood 79:3036–3044

Gieselmann V (1995) Lysosomal storage diseases. Biochim Biophys Acta 1270:103–126

Haylock DN, To LB, Dowse TL, Juttner CA, Simmons PJ (1992) Ex vivo expansion and maturation of peripheral blood CD34+ cells and unseparated paripheral blood progenitor cells. Blood 80:1405–1412

Henschler R, Brugger W, Luft T, Frey T, Mertelsmann R, Kanz L (1994) Maintenance of transplantation potential in ex vivo expanded CD34+ peripheral blood progenitor cells. Blood 84:2898–2903

Kessinger A, Armitage JO (1991) The evolving role of autologous peripheral stem cell transplantation following high-dose therapy for malignancies. Blood 77:211

Koller MR, Bender JG, Miller WM, Papoutsakis ET (1993a) Expansion of primitive human hematopoietic progenitors in perfusion bioreactor system with Il-3 Il-6 and stem cell factor. Biotechnology 11:358–363

Koller MR, Emerson SG, Palsson BO (1993b) Large-scale expansion of human stem and progenitor cells from bone marrow mononuclear cells in perfusion cultures. Blood 82 378–384

Koller MR, Palsson MA, Manchel I, Palsson BO (1995) Long-term culture initiating cell expansion is dependent on frequent medium exchange combined with stromal and other accessory cell effects. Blood 86:1784–1793

Lansdorp PM, Dragowska W (1993) Maintenance of hematopoiesis in serum free bone marrow cultures involves sequential recruitment of quiescent progenitors. Exp Hematol 21:1321–1327

Lansdorp PM, Dragowska W, Mayani H (1993) Ontogeny-related changes in proliferative potential of human hematopoietic cells. J Exp Med 178:787–791

Lieschke G, Burgess AW (1992) Granulocyte colony-stimulating factor and granulocyte-macrophage colony-stimulating factor. N Engl J Med 327:28–35, 99–106

Mayani H, Dragowska W, Lansdorp PM (1993a) Characterization of functionally distinct subpopulations of CD34+ cord blood cells in serum-free long term cultures supplemented with hematopoietic cytokines. Blood 82:2664–2672

Mayani H, Dragowska W, Lansdorp PM (1993b) Cytokine-induced selective expansion and maturation of erythroid versus myeloid progenitors from purified cord blood precursor cells. Blood 81:3252–3258

Maze R, Carney JP, Kelley MR, Glassner BJ, Williams DA, Samson L (1996) Increasing DNA repair methyltransferase levels via bone marrow stem cell transduction rescues mice from the toxic effects of 1,3-bis(2-chloroethyl)-1-nitrosourea, a chemotherapeutic alkylating agent. Proc Natl Acad Sci USA 92:206–210

Moore MAS (1991) Clinical implications of positive and negative hematopoietic stem cell regulators. Blood 78:1–19

Muench MO, Firpo MT, Moore MAS (1993) Bone marrow transplanation with interleukin-1 plus kit-ligand ex vivo expanded bone marrow accelerates hematopoietic reconstitution in mice without loss of stem cell lineage and proliferative potential. Blood 81:3463–3473

Naughton BA, Tjota A, Sibanda B, Naughton GK (1991) Hematopoiesis on suspended nylon screen-stromal cell microenvironments. J Biomech Eng 113:171–177

Nolta JA, Smogorzewska EM, Kohn DB (1995) Analysis of optimal conditions for retroviral-mediated transduction of primitive human hematopoietic cells. Blood 86:101–105

Oh DJ, Koller MR, Palsson BO (1994) Frequent harvesting from perfused bone marrow cultures results in increased overall cell and progenitor expansion. Biotechnol Bioeng 44:609–616

Peters SO, Kittler ELW, Ramshaw HS, Quesenberry PJ (1996) Ex vivo expansion of murine marrow cells with interleukin-3 (IL-3), IL-6, IL-11, and stem cell factor leads to impaired engraftment in irradiated hosts. Blood 87:30–37

Peters SO, Kittler ELW, Ramshaw AS, Quesenberry PJ (1995) Murine marrow cells expanded in culture with IL-3, IL-6, IL-11 and SCF aquire an engraftment defect in normal hosts. Exp Hematol 23:461–466

Peters WP (1995) High dose chemotherapy with autologous bone marrow transplantation for the treatment of breast cancer. In: DeVita VTS, Rosenberg SA (eds) Important advances in oncology. Lippincott, Philadelphia, pp 215–230

Pettengell R, Gurney H, Radford JA et al (1992) Granulocyte colony-stimulating factor to prevent dose-limiting neutropenia in non-Hodgkin's lymphoma: a randomized controlled trial. Blood 80:1430–1436

Ploemacher RE, van der Sluijs JP, van Beurden CAJ, Baert MRM, Chan PL (1991) Use of limiting-dilution type long-term marrow cultures in frequency analysis of marrow-repopulating and spleen colony-forming heamtopoietic stem cells in the mouse. Blood 78:2527–2533

Rebel VI, Dragowska W, Eaves CJ, Humphries RK, Lansdorp PM (1994) Amplification of Sca-1+ Lin- WGA+ cells in serum-free cultures containing steel factor, interleukin-6, and erythropoietin with maintenance of cells with long-term in vivo reconstituting potential. Blood 83:128–136

Ross AA, Cooper BW, Lazarus HM et al (1993) Detection and viability of tumor cells in peripheral blood stem cell collections from breast cancer patients using immunocytochemical and clonogenic assay techniques. Blood 82:2605–2610

Sardonini CA, Wu YJ (1993) Expansion and differentiation of human hematopoietic cells from static cultures through small-scale biorectors. Biotechnol Prog 9:131–137

Schwartz RM, Emerson SG, Clarke MF, Palsson BO (1991) In vitro myelopoiesis stimulated by rapid medium exchange and supplementation with hematopoietic growth factors. Blood 78:3155–3161

Sharp J, Kessinger A, Armitage JO, Biermann P, Crouse D, Mann S, Pirruccello S, Vose J, Weisenburger DD (1991) Clinical significance of occult tumor cell contamination of hematopoietic harvests in non-Hodgkin's lymphoma and Hogkin's disease. Proceedings of the international symposium on ABMT in lymphoma, Hodgkin's disease and multiple myeloma. Wilsede, Germany

Sharp JG, Kessinger A, Vaughan WP et al (1992) Detection and clinical significance of minimal tumor cell contamination of peripheral blood stem cell harvests. Int J Cell Cloning 10:92

Sheridan WP, Begley G, Juttner CA et al (1992) Effect of peripheral-blood progenitor cells mobilised by filgrastim (G-CSF) on platelet recovery after high-dose chemotherapy. Lancet 339 640–644

Srour EF, Brandt JE, Bridell RA, Gringsby S, Leemhuis T, Hoffman R (1993) Long term generation and expansion of human primitive hematopoietic progenitor cells in vitro. Blood 81:661–669

Testa NG, Dexter TM (1991) The biology of long-term bone marrow cultures and its application to bone marrow transplantation. Curr Opin Oncol 3:272–278

To LB, Roberts MM, Haylock DN, Dyson PG, Branford AL, Thorp D, Ho JQK, Dart GW, Horvath N, Davy MLJ, Olweny CLM, Abdi E, Juttner CA (1992) Comparison of hematological recovery times and supportive care requirements of autologous recovery phase peripheral stem cell transplants autologous bone marrow transplants and allogeneic bone marrow transplants. Bone Marrow Transplant 9:277–284

Verfaillie CM, Catanzarro PM, Li WN (1994) Macrophage inflammatory protein 1-alpha, interleukin-3, and diffusible marrow stroma factors maintain human hematopoietic stem cells for at least eight weeks in vitro. J Exp Med 179:643–649

Wang TY, Brennan JK, Wu JHD (1995) Multilineage hematopoiesis in a three-dimensional murine long-term bone marrow culture. Exp Hematol 22:26–32

Witte O (1990) Steel locus defines new multipotent growth factor. Cell 63: 5–6

Zandstra PW, Cameron G, Eaves CJ, Piret JM (1994) Hematopoietic progenitor cell expansion in stirred bioreactors. Blood 84:498a (abstract)

4 Monocyte-Derived Cells in Adoptive Immunotherapy of Cancer: Facts and Perspectives

S.W. Krause, B. Hennemann, A. Konur, M. Kreutz,
and R. Andreesen

4.1 Introduction

The immune system is able to cope with infectious and malignant disease by a complicated network that includes innate and adaptive mechanisms. Within this network, cells of the mononuclear phagocyte system play a prominent role in innate immunity due to their capabilities for phagocytosis, cytotoxicity and production of cytokines and other secretory products. Furthermore, an important link to specific immunity lies in their involvement in the presentation of antigen to T cells.

4.2 Functional Consequences of Macrophage Differentiation

Based on our current understanding, most resident tissue macrophages (MACs) arise from blood monocytes (MOs) after a process of terminal differentiation, although some proliferation of MACs may occur in certain circumstances (van Furth 1989). During this maturation process cells increase in size and a change in function and phenotype occurs. A similar step of differentiation from MOs to MACs can be studied in vitro and has become an important tool to examine the properties of these cells (Andreesen et al. 1983b). Contaminating lymphocytes do not affect this differentiation process, as long as no lymphocyte-activating stimuli are present. Nevertheless, MAC maturation can be modulated by a number of different agents. For example, stimulation either of MOs themselves by interferon-γ (IFNγ) or endotoxin (lipopolysaccharide, LPS) or stimulation of contaminating lymphocytes will interrupt MAC maturation (Brugger et al. 1991).

In unstimulated serum-containing cultures, MO-derived cells resemble tissue MACs after about 1 week. These changes in size and shape are accompanied by a shift of antigenic phenotype and functional properties (Andreesen et al. 1990a; Scheibenbogen and Andreesen 1991), a number of examples are summarized in Table 1. Many results depicted in this table are derived from studies with "prototype" in vitro derived MACs, as described above, but several features were also confirmed with naturally occuring MACs, mostly alveolar MACs. Our group has developed MAC-specific antibodies to be used as tools to examine differences between MOs and MACs (Andreesen et al. 1986). Recently, the nature of the antigen detected by antibodies MAX.1 and MAX.11 was elucidated and thus the reactivity of these antibodies was linked to a cellular function: the antigen is identical to carboxipeptidase M, a membrane bound ectopeptidase originally described in the placenta (Rehli et al. 1995). This enzyme modifies a vast number of different substrates, with MACs expressing a higher specific activity than all other cells examined yet.

One functional difference between MOs and MACs is especially important for their role as host defense cells: MACs are more cytotoxic than MOs against a number of tumor targets (Andreesen et al. 1983a, 1988). In Fig. 1A, the increased cytotoxicity (as measured in a postlabeling assay) of MOs during their differentiation into MACs is shown.

Table 1. Differences between monocytes and macrophages

	MO	MAC
Surface antigens		
Carboxypepetidase M (MAX. 1/11)	(+)	++
MAX.3	(+)	++
Fcγ (CD16)	5–10%	+++
Transferrin receptor (CD71)	(+)	++
Mannose receptor	(+)	++
Cytokine secretion (induced by LPS)		
TNF	+	++
IL-6	++	+
IL-1	++	–
GM-CSF	+	++
Other functions		
Antigen presentation (primed T cells)	++	++
Antigen presentation (unprimed T cells)	+	–
Phagocytosis	+	++
Tumor cytotoxicity	(+)	++
Procoagulant activity	+	++

MO, monocytes; MAC, macrophages; TNF, tumor necrosis factor; IL, inter-leukin, GM-CSF, granulocyte macrophage colony-stimulating factor.

Cells were stimulated with IFNγ before interaction with tumor cells in this assay, but unstimulated cells also exhibit some cytotoxic effector function (data not shown here). The tumor cell line U937 was used as a target, although it should be mentioned that the sensitivity of different tumor targets to the unspecific cytotoxic effect of macrophages varies from extremely sensitive to not sensitive at all.

In Fig. 1B the capability of MOs and MACs to be used as effector cells for antibody-mediated cellular cytotoxicity (ADCC) is shown. It was already known that MACs acquire the low-affinity IgG receptor CD16 during maturation, whereas this receptor is only present on a small subpopulation of blood MOs. In parallel, MACs become much better effectors of ADCC (Munn and Cheung 1989). Of course, a specific antibody that detects the tumor target must be present in this situation.

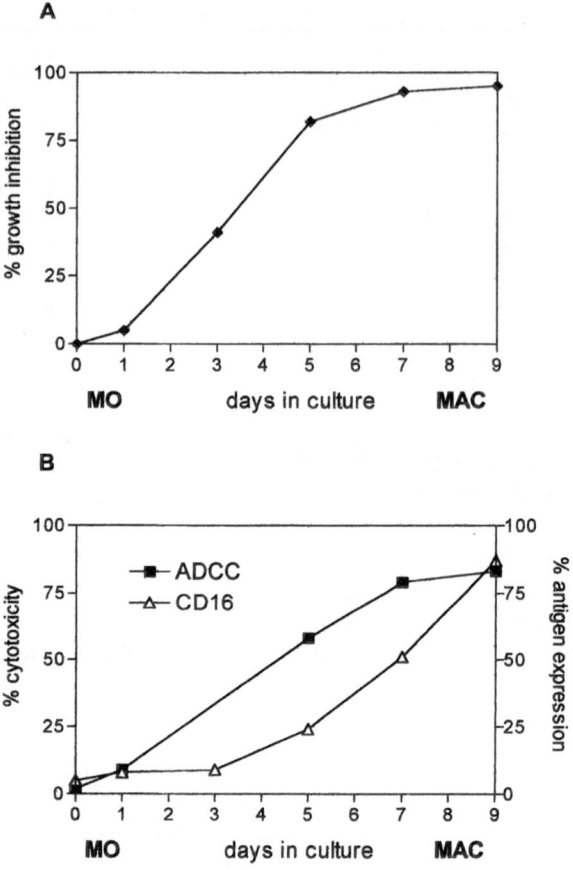

Fig. 1A, B. Tumor cytotoxicity during monocyte (*MO*) to (*MAC*) differentiation. MOs were allowed to differentiate into MACs in vitro. At different time
points MOs/MACs were harvested from the cultures and tested for cytotoxic
effector function. **A** Cytotoxity against cell line U937 was tested in a postlabeling assay. **B** Up-regulation of Fc receptor (FcR)-γIII (CD16) and activity in
an assay for antibody-dependent cytotoxicity (measured by cell-ELISA) are
depicted

4.3 Macrophage Functions Are Modulated by Tumor Cells

Unfortunately, tumor cytotoxicity is only one of many phenomena observed during the interaction between MACs and tumor cells. In vivo, MACs are present in many tumors and may constitute a prominent part of the inflammatory infiltrate. Mantovani et al. (1992) examined the features of such infiltrating MACs and coined the term "tumor-associated macrophages" (TAMs). TAMs are "deactivated" in many tumors; for example, production of oxygen radicals is reduced. There is even evidence suggesting that TAMs can help the tumor by supporting tissue invasion and neovascularization. From these data, a model of a "Janus-faced" MAC-tumor interaction was put forward, with the balance between "good" and "bad" properties of the MACs being shifted to one side or the other in any given tumor (Mantovani et al. 1992).

Whereas it was known that TAMs are different from other MACs in many circumstances, it was unclear, whether tumor cells are able to modulate MAC functions directly without the involvement of other mediator cells and which molecular mechanisms are involved. In order to gain further insight into MAC-tumor interaction we set up a three-dimensional in vitro system using multicellular spheroids as a model for a small tumor or micrometastasis in vivo (Sutherland 1988). If cells from many tumor cell lines are seeded onto an agarose surface, they stick to each other and form small balls of cells (spheroids). These spheroids continue to grow until a central necrosis develops. We used spheroids derived from cells of different bladder carcinoma lines and containing 40000 cells at the start of subsequent investigations.

In these coculture experiments we put MOs onto the spheroids and examined whether they would infiltrate these "microtumors". In fact, they did. After 2 h, MOs had clearly begun to migrate in between the tumor cells and after 24 h the whole spheroid was deeply infiltrated by MO (Fig. 2A). By contrast, MOs did not infiltrate spheroids derived from primary lines of normal skin fibroblasts (Konur et al. 1996).

As we thus observed that a close physical interaction between MOs and tumor cells was indeed taking place inside the spheroid cultures, we analyzed the differentiation of intraspheroidal MACs. Wheras MOs outside the spheroids strongly up-regulated the expression of maturation-associated antigens, this process was inhibited if MOs resided inside spheroids derived from the highly malignant bladder carcinoma

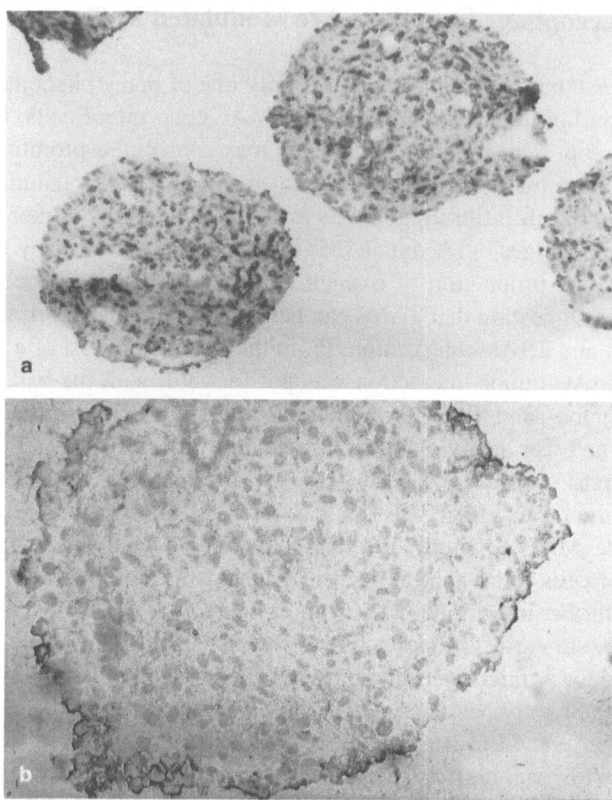

Fig. 2a,b. Monocytes/macrophages (MOs/MACs) inside spheroids of the bladder carcinoma line J82. MOs were allowed to infiltrate tumor cell spheroids and differentiate there for 1 week. **a** MOs are stained for the antigen CD11c present on all MOs and MACs. **b** Only MACs at the outer rim of the spheroid are stained for the maturation-associated marker carboxypeptidase M. The actual size of the spheroids is about 300 μm in **a** and **b**

Table 2. Antigen expression on monocytes afer 1 week coculture with spheroids of different bladder carcinoma cell lines

Antibody	Antigen expression on MO/MAC after coculture with spheroids			Control MAC
	J82	RT4	HCV29	from
CD11c	+	+	+	+
CD14	+	+	+	+
CD71	+	+	+	+
CD51	+	−	+	+
CD105	−	+	+	+
CPM	−	+	+	+
MAX.3	−	+	+	+

MO, monocytes; MAC, macrophages; CPM, carboxypeptidase M as determined by antibodies MAX.1 and MAX.11.
MO were cocultured with tumor spheroids for 1 week. Then antigen expression on infiltration MO/MAC was determined by immunohistochemistry.

cell line J82. Figure 2B shows that only MOs/MACs at the outer rim of the tumor spheroid stain positive for the MAC marker carboxypeptidase M, whereas the MOs/MACs inside the spheroid are not detected. Accordingly, suppression of this antigen does not occur in monolayer cocultures; instead, cell interaction inside these three-dimensional cultures seems to be necessary for this phenomenon. Parallel experiments were done for several MAC maturation-associated antigens and three different tumor cell lines. We found that the expression of several MAC antigens was suppressed by the least differentiated cell line J82. The better differentiated line RT4 suppressed the expression of one of the antigens examined, while the benign line HCV 29 did not modulate the antigens examined in these experiments. A summary of the results is given in Table 2. Interestingly, in vitro derived MAC also enter the spheroids, but their differentiation is not reversed inside: once maturation-associated antigens are expressed, they remain present on the cells (Konur et al. 1996).

The function of MOs interacting with tumor cells is also modulated. It was known that tumor cells can directly induce MOs or MACs to produce a variety of cytokines. What was new for us was the finding that MOs differentiating inside J82 spheroids keep a rather "monocytic"

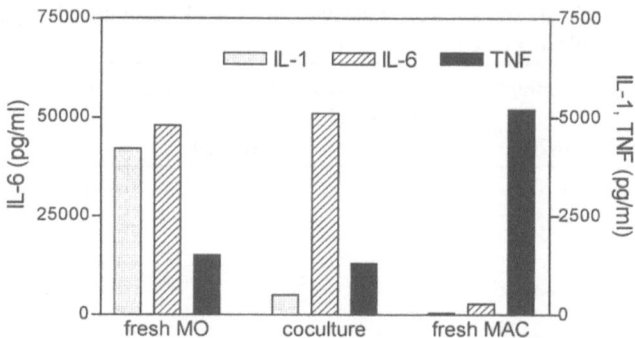

Fig. 3. Lipopolysaccharide (LPS)-induced cytokine production of mono-cytes/macrophages (*MO/MAC*) in tumor spheroid cocultures. MO were allowed to infiltrate J82 tumor cell spheroids and differentiate there for 1 week, then were stimulated with LPS. Cytokine content was measured in the super-natants (*coculture*). In control cultures either MO (*fresh MO*) or MAC harvested from Teflon cultures (*fresh MAC*) were cultured with J82 spheroids and stimulated with LPS at the initiation of the cocultures

cytokine secretion profile. MOs produce little tumor necrosis factor (TNF) but large amounts of interleukin-6 (IL-6) and IL-1, whereas mature MACs produce no IL-1, large amounts of TNF and little IL-6 (Scheibenbogen and Andreesen 1991). If MOs differentiate inside the J82 spheroid, they maintain secretion of IL-6, and to some degree of IL-1, but they do not up-regulate their TNF secretion (Fig. 3) (Konur et al. 1996). Moreover, if MACs are sorted from these spheroids, they still appear morphologically small compared to MOs. Taken together, it can clearly be shown that some tumor cells can suppress the normal development of mature MACs from MOs if three-dimensional cell interactions occur. The molecular mechanisms of this process are currently under study.

4.4 Adoptive Immunotherapy with Activated Macrophages

Since tumor cytotoxic MACs can be grown from blood MOs and since, by contrast, many tumors can modulate the function of TAMs to support tumor growth, it seemed logical to shift MAC differentiation such that development of "killer MACs" is enhanced while differentiation of "helper MACs" is reduced. Several trials have been reported in which MAC-activating agents, e.g., endotoxin, IFNγ or muramyldipeptide, were used in vivo for this purpose with limited succes (for a review see Hennemann and Andreesen 1996)

Another possible route to achieving this end is adoptive immunotherapy. Here, MOs are allowed to mature into MACs, with good anti-tumor effector function outside the patient in vitro, and are then reinfused. In animal models, this approach had been used with success (Bartholeyns et al. 1988). In our trials with cancer patients, white blood cells were harvested by leukopheresis and mononuclear cells were separated by Ficoll gradient centrifugation. Mononuclear cells containing about 20% MOs are then cultured for 1 week, during which MOs differentiated into mature MACs. Before reinfusion, MACs were separated from contaminating lymphocytes by counter-current elutriation and activated by additional stimuli.

Several routes of administration and different stimuli for activation of MACs were tested in Phase I studies. Five different protocols were carried out (Table 3). In the first three protocols, MACs were stimulated with IFNγ before reinfusion and three different routes of administration were explored: i.v. infusion, i.p. adminstration in patients with peritoneal carcinosis and intrahepatic administration via the hepatic artery in patients with prominent metastasis in the liver (Andreesen et al. 1990b; Hennemann et al. 1995). In the fourth protocol, MACs were stimulated with endotoxin before reinfusion, and in the fifth protocol MOs were mobilized by injection of GM-CSF before collection. Therapy was generally tolerated without major side effects. Low-grade fever and increases in C-reactive protein and circulating IL-6 could be observed in some patients, indicative of a biological response. Furthermore, the coagulation cascade was activated, evident from the elevation of fibrin monomers and thrombin-antithrombin complexes; yet, no resulting clinical side effects were noted (Andreesen et al. 1990b).

Table 3. Summary of adoptive immunotherapy trials with activated macrophages

	Intravenous (IFNγ[a])	Intraperitoneal (IFNγ)	Hepatic artery (IFNγ)	Intravenous (LPS)	Intravenous (GM-CSF mobilized)	Total
Number of patients in study	12	14	7	9	10	52
Therapy cycles	52	68	35	61	20	236
Maximum cell number/ therapy ($\times 10^8$)	17	10	13	15	27	
Maximum total cells	64	50	57	53	36	

IFN, interferon; GM-CSF, granulocyte macrophage colony-stimulating factor; LPS, lipopolysaccharide.
[a]Route of administration and activation stimulus of macrophages before reinfusion.

By the use of radioactively labeled MACs we could show, in selected patients, that an enrichment of activity occurred at the tumor site (Hennemann et al. 1995). The best therapeutic effects were noted when activated MACs were administered intraperitoneally in patients with peritoneal carcinosis. Here we saw a significant reduction of ascites, improvement of general well-being and decreased levels of tumor markers in several patients (Andreesen et al. 1990b). However, as no significant tumor reduction was achieved in these patients with end-stage disease, this therapy is no longer in use as a single treatment modality in this situation. Nonetheless, since some effects were observed, two other scenarios should be explored: (1) adoptive therapy of this kind could be applied in minimal residual disease or (2) it could be combined with other immunotherapy appoaches or with conventional therapy.

4.5 Myelomonocytic Cells in Antigen Presentation

In addition to their very important role in innate immunity, cells of the mononuclear phagocyte system are also play a critical role in specific immunity, a property which could also be exploited in therapeutic interventions. MOs or MO-derived antigen-presenting cells activate specific immunity, namely, the recruitment and activation of T lymphocytes, to generate a specific anti-tumor immune response. Furthermore,

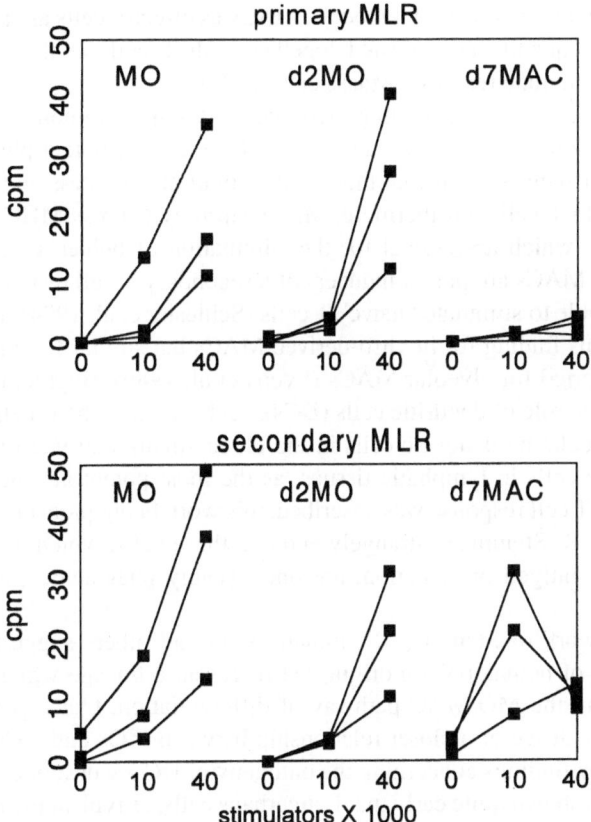

Fig. 4. Primary and secondary T cell stimulation by monocytes (*MO*) and macrophages (*MAC*). Either freshly isolated MO or MO differentiated for 2 days (*d2MO*) or 7 days (*7dMAC*) were used as stimulator cells in an allogeneic mixed lymphocyte reaction (*MLR*). Either freshly isolated lymphocytes (*primary MLR*) or lymphocytes harvested from previous stimulation in a first MLR with cells from the same donor (*secondary MLR*) were used as responder cells. Three experiments are shown

as shown in several animal models, MACs as effector cells are recruited and subsequently co-activated together with T cells at the site of a specific immune reaction (Maass et al. 1995).

In earlier work it was suggested that MACs are commonly involved in antigen presentation (Unanue 1984). Due to their potent phagocytic ability, it seemed clear that these cells would also process and present antigen to T cells. Furthermore, MACs strongly express MHC class II antigens, which are crucial for the stimulation of helper T cells. But whereas MACs are potent inducers of a secondary T cell response, they are not able to stimulate "naive" T cells (Schlesier et al. 1994). Figure 4 shows this finding for in vitro-derived MACs but similar findings were also reported for alveolar MACs (Lyons et al. 1986). Beginning in the 1970s, the role of dendritic cells (DCs), such as Langerhans cells of the skin, circulating dendritic cells in the blood stream and interdigitating dendritic cells in lymphatic tissues, as the most potent activators of a primary T cell response was described, this work being pioneered by the group of R. Steinman. Strangely enough, these cells, which are highly adept at antigen presentation, are only weakly phagocytic (Steinman 1991).

The work of Steinman (Steinman 1991) and others suggested that DCs are of hematopoietic origin, yet represent a lineage which is distinct from the MO/MAC pathway of differentiation. Other groups reported evidence of a closer relationship between MOs and DCs. Some of the key findings concerning the ontogeny of DC are described below:

It was shown quite early that Langerhans cells, as typical members of the DC family, are descendants of the bone marrow in murine chimeras as well as in humans after allogeneic bone marrow transplantation (Volc Platzer et al. 1984). Much later, it was demonstrated that typical DCs can be derived in vitro from CD34+ hematopoietic progenitor cells (Reid et al. 1990, 1992; Caux et al. 1992). A close relationship between the DC and MAC lineage of differentiation was suggested by the existence of mixed DC-MAC colonies arising from single CD34+ cells (Reid et al. 1992). Meanwhile, "veiled cells" with strong antigen-presenting cell function had been obtained in vitro from blood MOs, and a close relationship between these veiled cells and DCs was suggested (Peters et al. 1987). In 1994, evidence that DCs arise from peripheral blood precursor cells was published by two groups (Romani et al. 1994; Sallusto and Lanzavecchia 1994). The starting cell population for these

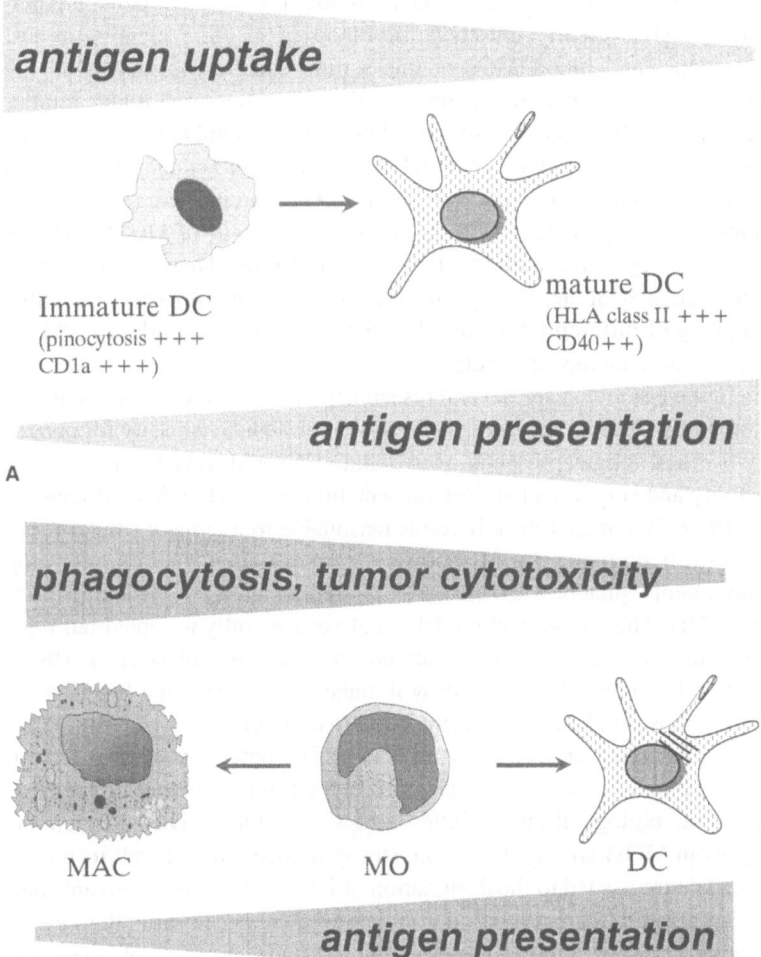

Fig. 5A, B. Two models of dendritic cell (*DC*) differentiation complementing each other. **A** Differentiation of immature into mature DCs. **B** Differentiation of monocytes (*MO*s) either into phagocytic mcrophages (*MAC*s) or antigen-presenting DCs

experiments was adherent peripheral blood mononuclear cells, a population that is clearly enriched for blood MOs. The identity of DC precursors and blood MOs in this setting can be assumed from the experiments but was not addressed directly. However, under similar culture conditions, cells with typical DC features could also be obtained from highly purified CD14$^+$ blood MOs (Zhou and Tedder 1996). Taken together, it remains unclear whether most DCs in vivo differentiate from bone marrow precursors via the intermediate form of blood MOs, as suggested by some authors (Peters et al. 1996). There is conclusive evidence that at least a special type of DC can proceed along this pathway of differentiation and therefore originates from the MO/MAC lineage of hematopoietic cells.

In the last few years, several researchers have reported terminal maturation of DCs from immature cells with intermediate capacity for phagocytosis and antigen presentation to mature DCs with very low phagocytic activity and very potent antigen presentation ability (Fig. 5A) (Romani et al. 1989; Steinman 1991). It seems reasonable to assume a similar process of differentiation for MOs as precursor cells that can either develop into mainly phagocytic cells or into mainly antigen-presenting cells (Fig. 5B). This theoretical model is not yet generally accepted but logically fits into the concept of antigen presentation (Peters et al. 1996). From a teleological point of view it makes sense, since the functions of phagocytosis and antigen presentation are reunited.

As discussed above, many aspects of DC ontogeny in vivo remain elusive and hopefully will be clarified in the future. However, disregarding some biological uncertainties, all DCs (including DCs differentiating from MOs) are, by definition, strong inducers of a T cell response. This feature has led to the designation of DCs as "nature's adjuvant" and has generated interest regarding their use in cancer immunotherapy. In several animal studies it has been shown that activation of a specific immune response can lead to tumor immunity and even to tumor regression (Pardoll 1992), and in some model systems DCs were used to induce such an effective immune response (Grabbe et al. 1995). Yet a note of caution is necessary, particularly considering the fact that not all DC populations derived from in vivo sources or differentiated from precursors in vitro will behave in the same manner. A variety of functions can be defined in different DC populations and it has to be assumed that different DC populations will induce different T cell

Table 4. Heterogenity of dendritic cell populations

Different cell sources
    Dendritic cells from in vivo sources (blood, skin, etc.)
    Preparation in vitro
        From monocytes (which conditions?)
        From CD34+ cells (which conditions?)

Different functional features
    Mobility
    Phagocytosis
    Fluid phase endocytosis
    Accessory molecules
    Cytokine repertoire
    Antigen presentation with MHC class II
    Antigen presentation with MHC class I

Will different dendritic cell populations lead to different types
of T cell respronses?
    Quantitative differences
    T helper 1 vs T helper 2
    Activation of cytotoxic T cells
    Preference for different antigens
    Anergy?

responses, at least quantitatively, but probably also qualitatively (Table 4).

Of course, the properties of different DC populations, such as the ability to migrate and interact with various cell populations, will have more or less impact depending on the particular immune therapy scenario. The ability to phagocytose and process antigen will be more important for native antigens than for synthetic peptides. Other features of the cells will lead to effects that can only partially be deduced theoretically and thus have to be determined practically. We have started to compare different populations of antigen-presenting cells derived from MOs in vitro and have included DCs directly differentiated from CD34[+] progenitor cells. It turned out that DCs with a quite similar morphology and antigen profile can be distinguished from progenitor cells and from MOs. Functionally, differences do exist; for example,

CD34+-derived DCs are the strongest inducers of an allogeneic mixed lymphocyte reaction.

Recently, the first DC-based clinical trial was reported for patients with follicular lymphoma (Hsu et al. 1996). Four patients were treated in this study. The specific idiotype immunoglobulin of each patient was used as a tumor antigen and produced by fusion of lymphoma cells with immortal cell lines. Primary autologous DCs were separated from the peripheral blood by a multistep process and used for immunization after cultivation with the target antigen. In three patients a clinical tumor regression was observed indicating that DC-based adoptive immunotherapy to induce specific tumor immunity can be successful. However, the methods used in this trial are too complicated and too expensive to be used for treating a large number of patients.

Many questions remain: Can other tumor antigens or even whole tumor cells be used? Will in vitro-derived DCs (much easier to generate, especially MO-derived DCs) be as effective as primary DCs? Which type of DCs will work best in which immunization schedule? Will DCs be better than genetically modified tumor cells or even traditional optimized adjuvants? Will there be a place for the additional activation of MACs? Hopefully, ongoing studies will soon tell us whether specific adoptive immunotherapy for the treatment of cancer will benefit a larger population of patients.

References

Andreesen R, Osterholz J, Bross KJ, Schulz A, Luckenbach GA, Lohr GW (1983a) Cytotoxic effector cell function at different stages of human monocyte-macrophage maturation. Cancer Res 43: 5931–5936

Andreesen R, Picht J, Lohr GW (1983b) Primary cultures of human blood-born macrophages grown on hydrophobic teflon membranes. J Immunol Methods 56: 295–304

Andreesen R, Bross KJ, Osterholz J, Emmrich F (1986) Human macrophage maturation and heterogeneity: analysis with a newly generated set of monoclonal antibodies to differentiation antigens. Blood 67: 1257–1264

Andreesen R, Gadd S, Brugger W, Lohr GW, Atkins RC (1988) Activation of human monocyte-derived macrophages cultured on Teflon: response to interferon-gamma during terminal maturation in vitro. Immunobiology 177: 186–198

Andreesen R, Brugger W, Scheibenbogen C, Kreutz M, Leser HG, Rehm A, Lohr GW (1990a) Surface phenotype analysis of human monocyte to macrophage maturation. J Leukoc Biol 47: 490–497

Andreesen R, Scheibenbogen C, Brugger W, Krause S, Meerpohl HG, Leser HG, Engler H, Lohr GW (1990b) Adoptive transfer of tumor cytotoxic macrophages generated in vitro from circulating blood monocytes: a new approach to cancer immunotherapy. Cancer Res 50: 7450–7456

Bartholeyns J, Lombard Y, Dumont S, Hartmann D, Chokry M, Giaimis J, Kaufmann S, Poindron P (1988) Immunotherapy of cancer: experimental approach with activated macrophages proliferating in culture. Cancer Detect Prev 12: 413–420

Brugger W, Reinhardt D, Galanos C, Andreesen R (1991) Inhibition of in vitro differentiation of human monocytes to macrophages by lipopolysaccharides (LPS): phenotypic and functional analysis. Int Immunol 3: 221–227

Caux C, Dezutter Dambuyant C, Schmitt D, Banchereau J (1992) GM-CSF and TNF-alpha cooperate in the generation of dendritic Langerhans cells. Nature 360: 258–261

Grabbe S, Beissert S, Schwarz T, Granstein RD (1995) Dendritic cells as initiators of tumor immune responses: a possible strategy for tumor immunotherapy? Immunol Today 16: 117–121

Hennemann B, Scheibenbogen C, Schumichen C, Andreesen R (1995) Intrahepatic adoptive immunotherapy with autologous tumorcytotoxic macrophages in patients with cancer. J Immunother 18: 19–27

Hennemann B Andreesen R (1996) Monocyte/macrophage activation by immunostimulators – role in cancer therapy. Clin Immunother 5: 294–308

Hsu FJ, Benike C, Fagnoni F, Liles TM, Czerwinski D, Taidi B, Engleman EG, Levy R (1996) Vaccination of patients with B-cell lymphoma using autologous antigen-pulsed dendritic cells. Nat Med 2: 52–58

Konur A, Kreutz M, Knuchel R, Krause SW, Andreesen R (1996) Three-dimensional coculture of human monocytes and macrophages with tumor cells-analysis of macrophage differentiation and activation. Int J Cancer 66: 645–652

Lyons CR, Ball EJ, Toews GB, Weissler JC, Stastny P, Lipscomb MF (1986) Inability of human alveolar macrophages to stimulate resting T cells correlates with decreased antigen-specific T cell-macrophage binding. J Immunol 137: 1173–1180

Maass G, Schmidt W, Berger M, Schilcher F, Koszik F, Schneeberger A, Stingl G, Birnstiel ML, Schweighoffer T (1995) Priming of tumor-specific T cells in the draining lymph nodes after immunization with interleukin 2-secreting tumor cells: three consecutive stages may be required for successful tumor vaccination. Proc Natl Acad Sci U S A 92: 5540–5544

Mantovani A, Bottazzi B, Colotta F, Sozzani S, Ruco L (1992) The origin and function of tumor-associated macrophages. Immunol Today 13: 265–270

Munn DH Cheung NK (1989) Antibody-dependent antitumor cytotoxicity by human monocytes cultured with recombinant macrophage colony-stimulating factor. Induction of efficient antibody-mediated antitumor cytotoxicity not detected by isotope release assays. J Exp Med 170: 511–526

Pardoll D (1992) New strategies for active immunotherapy with genetically engineered tumor cells. Curr Opin Immunol 4: 619–623

Peters JH, Ruhl S, Friedrichs D (1987) Veiled accessory cells deduced from monocytes. Immunobiology 176: 154–166

Peters JH, Gieseler R, Thiele B, Steinbach F (1996) Dendritic cells – from ontogenetic orphans to myelomonocytic descendants. Immunol Today 17: 273–278

Rehli M, Krause SW, Kreutz M, Andreesen R (1995) Carboxypeptidase M is identical to the MAX.1 antigen and its expression is associated with monocyte to macrophage differentiation. J Biol Chem 270: 15644–15649

Reid CD, Fryer PR, Clifford C, Kirk A, Tikerpae J, Knight SC (1990) Identification of hematopoietic progenitors of macrophages and dendritic Langerhans cells (DL-CFU) in human bone marrow and peripheral blood. Blood 76: 1139–1149

Reid CD, Stackpoole A, Meager A, Tikerpae J (1992) Interactions of tumor necrosis factor with granulocyte-macrophage colony-stimulating factor and other cytokines in the regulation of dendritic cell growth in vitro from early bipotent CD34+ progenitors in human bone marrow. J Immunol 149: 2681–2688

Romani N, Lenz A, Glassel H, Stossel H, Stanzl U, Majdic O, Fritsch P, Schuler G (1989) Cultured human Langerhans cells resemble lymphoid dendritic cells in phenotype and function. J Invest Dermatol 93: 600–609

Romani N, Gruner S, Brang D, Kampgen E, Lenz A, Trockenbacher B, Konwalinka G, Fritsch PO, Steinman RM, Schuler G (1994) Proliferating dendritic cell progenitors in human blood. J Exp Med 180: 83–93

Sallusto F Lanzavecchia A (1994) Efficient presentation of soluble antigen by cultured human dendritic cells is maintained by granulocyte/macrophage colony-stimulating factor plus interleukin 4 and downregulated by tumor necrosis factor alpha. J Exp Med 179: 1109–1118

Scheibenbogen C Andreesen R (1991) Developmental regulation of the cytokine repertoire in human macrophages: IL-1, IL-6, TNF-alpha, and M-CSF. J Leukoc Biol 50: 35–42

Schlesier M, Krause S, Drager R, Wolff Vorbeck G, Kreutz M, Andreesen R, Peter HH (1994) Monocyte differentiation and accessory function: different effects on the proliferative responses of an autoreactive T cell clone as compared to alloreactive or antigen-specific T cell lines and primary mixed lymphocyte cultures. Immunobiology 190: 164–174

Steinman RM (1991) The dendritic cell system and its role in immunogenicity. Annu Rev Immunol 9: 271–296

Sutherland RM (1988) Cell and environment interactions in tumor microregions: the multicell spheroid model. Science 240: 177–184

Unanue ER (1984) Antigen-presenting function of the macrophage. Annu Rev Immunol 2: 395–428

van Furth R (1989) Origin and turnover of monocytes and macrophages. Curr Top Pathol 79: 125–150

Volc Platzer B, Stingl G, Wolff K, Hinterberg W, Schnedl W (1984) Cytogenetic identification of allogeneic epidermal Langerhans cells in a bone-marrow-graft recipient. N Engl J Med 310: 1123–1124

Zhou LJ, Tedder TF (1996) CD14+ blood monocytes can differentiate into functionally mature CD83+ dendritic cells. Proc Natl Acad Sci U S A 93: 2588–2592

5 The Immunological Basis
for the Development of Tumor Cell- and Peptide-Based Vaccines for Treatment of Patients with Renal Cell Carcinoma

D. J. Schendel, E. Nößner, B. Maget, S. Kressenstein, K. Pantel, and R. Oberneder

5.1 Introduction

The failure of metastatic renal cell carcinoma (RCC) to respond to irradiation, chemotherapy or hormone therapy has left immunotherapy as the most hopeful form of treatment to date. Systemic cytokine therapy of RCC using recombinant interleukin-2 (rIL-2) alone or in combination with recombinant interferon-α (rIFN-α) has led to objective response rates of 20%–30% (Belldegrun et al. 1991; Rosenberg 1992). Nevertheless, few complete and long-lasting remissions have been achieved and severe toxic side effects limit their broad application. These obstacles have spurred efforts to develop alternative strategies to modulate immune responses of RCC patients. In particular, new approaches are sought for treatment of patients with minimal residual disease. In the forefront are attempts to develop genetically engineered tumor cell vaccines and peptide vaccines that will induce tumor-specific immune responses in vivo.

Analysis of anti-tumor responses of patients with melanoma serve as a paradigm for studies of other solid tumors. In the 10 years since the first identification was made of specific cytotoxic T lymphocytes (CTLs) that were able to eliminate autologous tumor cells, a variety of immunotherapies have been developed for treatment of metastatic melanoma (reviewed in Parmiani 1990; Knuth et al. 1991; Oettgen 1991; Boon et al. 1994). These include among others: (1) systemic application of cytokines to enhance immune responses in vivo, (2) adoptive transfer of lymphokine activated killer (LAK) cells or tumor-infiltrating lymphocytes (TILs) following activation and expansion in vitro and, (3) vaccination with autologous or allogeneic tumor cells. More recently, modifications of these strategies have been initiated using gene transfer methods. Clinical studies have been initiated to analyze the effect of introducing cytokine genes, such as tumor necrosis factor (TNF) or rIL-2, into immune effector cells to enhance their anti-tumor activity. Autologous or allogeneic tumor cells have been genetically engineered to secrete various cytokines with the goal to improve their ability to activate immune responses in vivo. Likewise, a major histocompatibility complex (MHC) gene has been delivered to tumor cells in vivo to stimulate allogeneic responses similar to those causing transplant rejection. The recruitment of alloreactive lymphocytes and their secretion of complex mixtures of cytokines appear to allow the development of effector cells that can simultaneously destroy nearby unmodified tumor cells (reviewed in Rosenberg 1992; Parmiani 1990; Knuth et al. 1991; Oettgen 1991; Boon et al. 1994). In addition, animal studies have shown that improved generation of tumor-specific CTLs can occur in vivo in the environment of an active allogeneic response (Mandelboim et al. 1995). Recent developments have allowed the molecular identification of several MHC class I/peptide complexes that are expressed by melanoma cells and recognized by tumor-associated CTLs (Boon et al. 1994; Finn 1993). This has opened the way to develop synthetic peptide vaccines. Initial clinical observations of peptide vaccination of melanoma patients with advanced disease reveal an important new potential for patient treatment (Marchand et al. 1995; Jaeger et al. 1996).

5.2 Immune Responses to Renal Cell Carcinomas

Application of new immunotherapies for other solid tumors, such as RCC, lag in development because the precise cellular and molecular basis of anti-tumor immune responses is less well characterized. Several observations suggest that some RCC can elicit immune responses: spontaneous tumor regression has been observed occasionally and cytokine-induced remission has also been achieved in some patients. Lymphocytic infiltrates accumulate in or near primary and metastatic lesions; when infiltrating lymphocytes were isolated and cultured in vitro they often displayed anti-tumor activity (reviewed in Schendel et al. 1997).

This knowledge was judged sufficient to initiate clinical trials of several new treatments for metastatic RCC, including the adoptive transfer of autologous, genetically modified effector cells to patients with advanced disease or the vaccination of patients with autologous or allogeneic genetically engineered tumor cells, in analogy to treatments designed for melanoma (Belldegrun et al. 1991; Rosenberg 1992; Parmiani 1990; Knuth et al. 1991; Oettgen 1991; Boon et al. 1994). However, this level of understanding of immune responses directed against RCC is clearly insufficient to evaluate the reasons for success or failure of treatment in individual patients. Here detailed information about the cellular mechanisms of anti-RCC responses is essential. Moreover, if new strategies such as peptide vaccination are to be developed for minimal residual disease, it is imperative to define the molecular basis of specific recognition of tumor cells (Nossner and Schendel 1997; Schendel et al. 1997).

We have used the following approach to examine immune responses directed against RCC (Schendel et al. 1993; Schendel and Gansbacher 1993). TILs were isolated from freshly resected tumors and cultured in vitro in the presence of autologous tumor cells and exogenous cytokines. The expanded populations were characterized with respect to function, specificity and T cell receptor (TCR) usage to define the molecular basis of interactions between tumor cells and lymphocytes in vitro. Their biological significance in vivo was assessed by determining the prevalence of specific T cells in tumor samples cryopreserved at the time of surgery, using the unique TCR sequence of a given T cell clone as a molecular marker in situ. In several cases we were able to identify highly specific CD3+/CD8+ CTLs that lysed autologous and allogeneic

tumor cells in an MHC-restricted manner via recognition of tumor-associated peptide(s) presented by HLA-A2 (Schendel et al. 1993). Some of
these CTLs belonged to dominant families of T cells that had accumu-
lated at the tumor site (Jantzer et al. 1995; and Jantzer and Schendel,
1997), suggesting that they are components of a complex immune
response that was initiated in vivo.

Several important parameters regarding the nature of RCC-associated immune responses could be defined once such specific CTLs were
obtained. First, the cytotoxic activity of these CTLs was tumor-specific
since lysis of class I-negative target cells like Daudi and K562 was not
observed, nor were corresponding autologous normal kidney cell
(NKC) lines recognized (Fig. 1A, B). Second, a permanent tumor cell
line (RCC-26) was established from patient 26 and found to be a
suitable target cell for CTL populations isolated from several other
HLA-A2-positive patients (Fig. 1B). Thus, RCC-26 tumor cells express
one or more common tumor-associated ligands that are presented by

---

Fig. 1A–D. Cytotoxic T lymphocyte (CTL) responses directed against renal ▶
cell carcinomas (*RCC*). **A** The tumor infiltrating lymphocytes (*TIL*) of RCC
patients 22, 26 and 29 were found to contain tumor-specific CTLs that lysed
autologous tumor but not autologous normal kidney cells (*NKC*) and the class
I negative target cell K562, which is particularly sensitive to human natural killer cells. **B** CTLs present in TIL of patients 22, 29 and 53 showed lysis of the
allogeneic tumor line established from patient 26 (*RCC-26*). The only HLA
class I molecule shared by all of these individuals was HLA-A2. As with
K562, lysis of the class I-negative cell line Daudi, which is sensitive to lymphokine activated killer cells, was not observed. **C** The autologous TIL (*TIL-
26*) population of patient 26 was found to show enhanced recognition of
autologous tumor cells genetically engineered to secrete interleukin-2 (*IL-2*)
(Schendel and Gansbacher 1993). Additional enhancement in recognition was
seen when RCC-26 and RCC-26/IL-2 cells were pretreated with interferon-α
(*IFNα*). **D** The CTLs present in the TIL population of patient 53 (*TIL-53*) were
not able to recognize autologous tumor cells (*RCC-53*) following restimulation
with these cells in the presence of exogenous cytokine, but showed weak lysis
of RCC-26 cells. When TIL-53 were restimulated with RCC-26/IL-2 cells
their cytotoxic potential against unmodified RCC-26 cells was enhanced. Nevertheless, weak but significant lysis of autologous RCC-53 cells only occurred
when they were pretreated with IFNα. The CTLs were tested at effector to target cell ratios ranging from 2:1 to 8:1 in a standard 4 h chromium release assay. Data are presented as percentage lysis or relative cytotoxic response (%
RCR) calculated as described previously (Schendel et al. 1979)

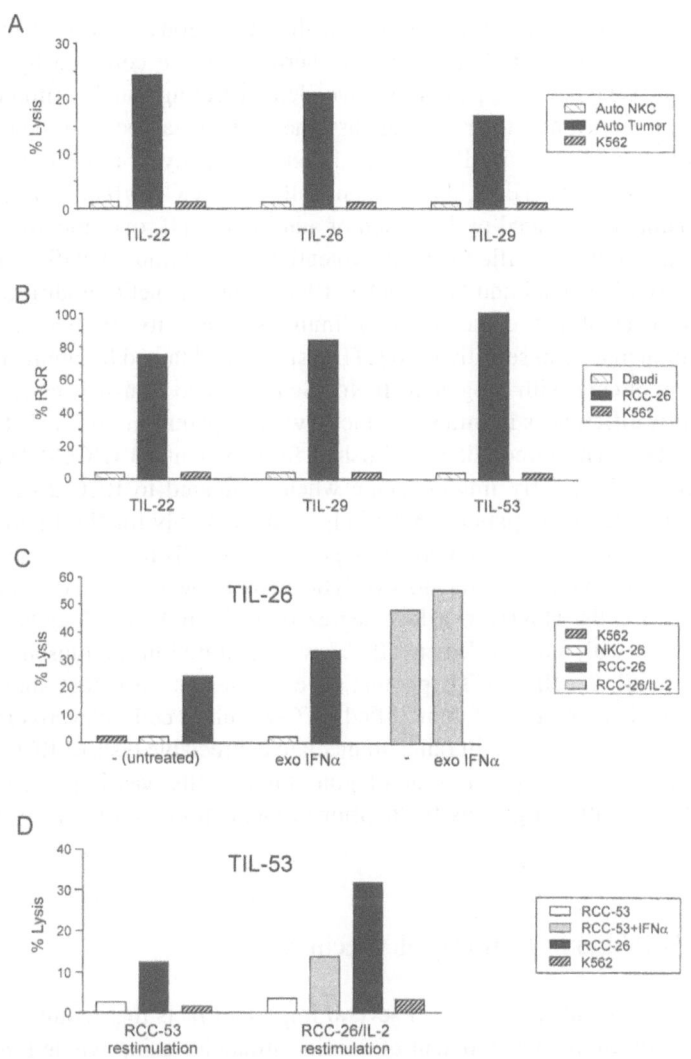

Fig. 1A–D. Legend see p. 78

HLA-A2 molecules and shared with the autologous tumors of other HLA-A2-positive RCC patients. Furthermore, these common ligands induced CTLs in these patients in vivo, demonstrating that their immune systems were not tolerized against these tumor-associated ligands. Third, when RCC-26 cells were modified genetically by retroviral gene transfer to secrete rIL-2, the resultant cell line (RCC-26/IL-2) was able to maintain sustained proliferation of autologous TIL over months and to enhance the specific cytotoxic potential of the tumor-specific CTLs (Fig. 1C; Schendel and Gansbacher 1993). Thus, genetic modification could even improve the immunostimulatory capacity of these cells. Small increases in sensitivity to CTL lysis were obtained by pretreating the tumor cells with exogenous IFNα (Schendel and Gansbacher 1993). Finally, differences in immunogenicity were apparent in different RCC (Fig. 1D). The tumor line established from patient 53 (RCC-53) appeared to be poorly immunogenic when compared to RCC-26 cells. TILs isolated from patient 53 could be sustained only for short periods of time by restimulation with autologous tumor cells in the presence of exogenous cytokines and no cytolytic activity was detected against RCC-53 cells. However, a low degree of lysis of RCC-26 cells was observed with TIL-53. When TIL-53 were cultured in the presence of RCC-26/IL-2 cells a CTL population expanded in vitro that showed enhanced recognition of unmodified RCC-26 tumor cells. Nevertheless, these activated CTLs still failed to mediate appreciable lysis of RCC-53 tumor cells. Lysis of these autologous tumor cells was improved by treatment with exogenous IFNα prior to incubation with the activated CTLs.

5.3 Autologous Tumor Cell Vaccines

The experiments above reveal several important facts that require consideration in the development of vaccine strategies using whole tumor cells. Currently, two types of genetically engineered tumor cell vaccines are considered for RCC patients (Anonymous 1996; Nossner and Schendel 1997) (Table 1). The first involves the genetic modification of a patient's own tumor cells and is designated as "autologous vaccination." Hereby, a tumor is excised and tumor cells are genetically modified ex vivo to enhance their immunogenicity by introducing one or

Table 1. Comparision of tumor cell- and peptide based vaccine strategies for treatment of renal cell carcinoma

Autologous tumor cell vaccine	Allogenic tumor cell vaccine	Tumor-specific peptide vaccine
Each tumor must be genetically engineered	Only one tumor line must be genetically engineered	No genetic engineering of cells is required
Not all tumors may grow. Immunogenicity must be determined for each tumor	Highly immunogenic tumor lines can be selected	Immunogenic peptides can be selected and synthetically manufactured
Full MHC matching is given	MHC matching is limited. Allorecognition may enhance responses	MHC matching remains limited
Unique and common tumor ligands can targeted	Only common tumor ligands can be targeted	Only common tumor peptides can be targeted
Technically difficult	**Feasible today**	**Broad future application**

several immunostimulatory molecules, such as various cytokines and/or costimulatory molecules like B7.1 (Pardoll 1993, 1995; Allison et al. 1995). Modified tumor cells are then readministered to the same individual. If tumor cells are limiting or cannot be cultured to allow genetic modification, then other cells of a patient, such as fibroblasts, can be modified and admixed with unmodified tumor cells in the administered vaccine (Kim and Cohen 1994; Tahara et al. 1994; Veelken et al. 1994). Alternatively, modifying genes can be directly introduced into tumors in vivo if they are surgically accessible (Plautz et al. 1993; Hersh et al. 1994). Autologous vaccines are patient-specific and the same procedures are repeated to generate a unique vaccine for each individual patient. In autologous vaccines, peptides can be presented by any of the MHC molecules expressed by the tumor cells since lymphocyte subsets of the patient have been selected during thymic development to recognize peptides presented by each self-MHC molecule. This full MHC matching has the potential to allow multiple target ligands, including peptides derived from tumor-associated proteins expressed only by a patient's own tumor cells (i.e., unique ligands), as well as peptides derived from proteins common to different tumors to be utilized for immune stimulation.

In our experience two factors strongly limit this approach. First, only a low percentage of tumor cells proliferate adequately ex vivo to allow their genetic modification. Although fibroblast lines can be established routinely this requires extensive time, cost and effort and some modifications are not feasible. For example, the transfer of genes allowing expression of costimulatory molecules of the B7 family needs to be done directly into tumor cells since tumor-associated TCR ligands and costimulatory molecules must be presented by the same cell to enable individual lymphocytes to receive two activation signals simultaneously (Schwartz 1992). Second, extensive studies assessing tumor immunogenicity have to be made for each autologous tumor cell vaccine in order to avoid a situation such as that observed with RCC-53 tumor cells. The vaccine potential of such tumor cells may not be improved by genetic engineering because of their poor expression of TCR ligands. At this early stage of tumor cell vaccine development the failure to assess such parameters could easily lead to discouraging conclusions about the feasibility of using such approaches for treatment of some patients. These two handicaps seem to outweigh the potential benefit of autologous vaccines which provide full MHC matching with the T cell repertoire of a patient, thereby allowing both unique and common tumor-associated ligands to be targeted.

5.4 Allogenic Tumor Cell Vaccines

The second approach is that of "allogeneic tumor cell vaccination" (Table 1). In this case tumor cell lines established from one or several patients are used to vaccinate selected patients whose tumors have the potential to express common ligands presented by the vaccine cells. The major advantage of allogeneic vaccination is that well characterized, highly immunogenic tumor lines can be selected for vaccine development and their genetic engineering must be done only once. For example, a genetically engineered RCC-26 vaccine, with demonstrated high immunogenicity, could be utilized in the treatment of HLA-A2 matched patients such as RCC patients 22, 29 and 53. Approximately 50% of all RCC patients might be treated with a vaccine based on RCC-26 cells because of the high frequency of the HLA-A2 allele in the Caucasian population (Browning and Krausa 1996). However, even in this situation the status of the patient's tumor cells will have a strong impact on vaccine efficacy. In the case of patient 53 systemic administration of IFNα together with the allogeneic vaccine might be necessary to help activated lymphocytes to detect residual tumor cells. The disadvantages of allogeneic vaccination include the limitation in MHC matching that can be achieved between any patient and vaccine and the restricted targeting to common ligands. The use of small pools of genetically engineered tumor cells that express different MHC molecules, each of which presents common ligands, may help to overcome this handicap.

5.5 Peptide Vaccines

Tumor cell vaccines will find only limited application in specialized centers because of their technical complexity. Therefore, identification of tumor-associated peptides recognized by MHC-restricted CTL is critical for the development of peptide-based vaccines. Our results and those of others (Bernhard et al. 1994) demonstrate that such peptides must exist for RCC. The way to define these peptides has been paved by the efforts made over the past several years and the technical break-throughs regarding melanoma (Boon et al. 1994; Slinguff et al. 1994; Van Pel et al. 1995). Similar approaches may allow RCC-associated peptides to be defined in the future. The technical simplicity for making and delivering peptide vaccines clearly offers substantial advantages, enabling large numbers of cancer patients to receive treatment. If patients with minimal residual disease who are at high risk to relapse with metastases can be identified reliably, treatment with peptide vaccines may be particularly beneficial since these individuals have fewer tumor cells that must be destroyed and their immune systems are most capable of developing vigorous responses. These advantages merit the investment of extensive effort to identify the tumor-associated peptides recognized by RCC-specific CTLs.

5.6 Summary

Our systematic analyses of immune responses in patients with RCC have revealed that some tumors express antigenic determinants that allow them to be distinguished from their normal kidney cell counterparts. Most importantly, tumor-specific, MHC-restricted CTLs were identified in several patients. This indicates that immune responses against autologous tumor cells were initiated in vivo in these individuals even in the absence of any immunomodulatory treatment. Therefore, the immune system is fully able to respond to these spontaneously arising carcinomas. This suggests that vaccine strategies targeting RCC are feasible. In fact, in some patients de novo responses must not be induced. Rather, primed lymphocytes with tumor specificity may already exist. If appropriate means can be found to reactivate these cells through tumor cell or peptide vaccination, they may be particularly effective in finding and destroying residual tumor cells. It is also hoped that such vaccines will induce highly specific, long-lived memory CTLs in vivo that can hinder the outgrowth of micrometastases over years.

Acknowledgements. This work was supported by a grant of the Deutsche Forschungsgemeinschaft (Ho1596/3–1). We thank G. Riethmüller and A. Hofstetter for enthusiastic support of these studies.

References

Allison JP, Hurwitz AA, Leach DR (1995) Manipulation of costimulatory signals to enhance antitumor T-cell responses. Curr Opinion Immunol 7:682–686

Anonymous (1996) Clinical protocols: cytokine/immunotherapy. Cancer Gene Ther 3:58–68

Belldegrun A, Figlin R, Haas G, deKernion J (1991) Immunotherapy for metastic renal-cell carcinoma. World J Urol 9:157–159

Bernhard H, Karbach J, Wölfel T, et al. (1994) Cellular immune response to human renal-cell carcinomas: Definition of a common antigen recognized by HLA-A2-restricted cytotoxic T-lymphocyte (CTL) clones. Int J Cancer 59:837–842

Boon T, Cerottini JC, Van den Eynde B, Van der Bruggen P, Van Pel A (1994) Tumor antigens recognized by T lymphocytes. Ann Rev Immunol 12:337–365

Browning M, Krausa P (1996) Genetic diversity of HLA-A2: Evolutionary and functional significance. Immunol Today 17:165–170

Finn OJ (1993) Tumor-rejection antigens recognized by T lymphocytes. Curr Opin Immunol 5:701–708

Hersh EM, Akporiaye E, Harris D, et al. (1994) Phase I study of immunotherapy of malignant melanoma by direct gene transfer. Hum Gene Ther 5:1371–1384

Jaeger E, Bernhard H, Romero P et al. (1996) Generation of cytotoxic T-cell responses with synthetic melanoma-associated peptides in vivo: implications for tumor vaccines with melanoma-associated antigens. Int J Cancer 66:162–169

Jantzer P, Oberneder R, Maget B, Schendel DJ (1995) Recruitment of MHC-restricted cytotoxic T lymphocytes specific for renal cell carcinoma to the tumor in situ. In: Bukowski RM, Finke JH, Klein EA (eds) Biology of renal cell carcinoma. Springer, Berlin, Heidelberg, NewYork, pp 84–93

Jantzer P, Schendel DJ (1997) Immunosurveillance against spontaneously arising human renal cell carcinomas: antigen-driven selection in the T cell receptor repertoire of tumor-infiltrating lymphocytes in vivo. Submitted.

Kim TS, Cohen EP (1994) Immunization of mice with allogeneic fibroblasts genetically modified for interleukin-2-secretion and expression of melanoma-associated antigens stimulate predetermined classes of anti-melanoma effector cells. J Immunother 16:24–35

Knuth A, Wölfel T, Meyer zum Büschenfelde KH (1991) Cellular and humoral immune responses against cancer: implications for cancer vaccines. Curr Opin Immunol 3:659–664

Mandelboim O, Vadai E, Feldman M, Eisenbach L (1995) Expression of two H-2 K genes, syngeneic and allogeneic, as a strategy for potentiating immune recognition of tumor cells. Gene Therapy 2:757–765

Marchand MP, Weynants P, Rankin E, et al. (1995) Tumor regression responses in melanoma patients treated with a peptide encoded by gene MAGE-3. Int J Cancer 63:883–885

Nossner E, Schendel DJ (1997) Autologous and allogeneic tumor cell vaccines. In: Blankenstein T, Herrmann F (eds) Gene therapy: principles and applications. Chapman and Hall, Weinheim

Oettgen HF (1991) Cytokines in clinical cancer therapy. Curr Opin Immunol 3:699–705

Pardoll DM (1993) New strategies for enhancing the immunogenicity of tumors. Curr Opin Immunol 5:719–725

Pardoll DM (1995) Paracrine cytokine adjuvants in cancer immunotherapy. Ann Rev Immunol 13:399–415

Parmiani G (1990) An explanation of the variable clinical response to interleukin2 and LAK cells. Immunol Today 11:113–115

Plautz GE, Yang ZY, Wu BY, et al. (1993) Immunotherapy of malignancy by in vivo gene transfer into tumors. Proc Natl Acad Sci USA 90:4645–4649

Rosenberg SA (1992) Karnofsky Memorial Lecture. The immunotherapy and gene therapy of cancer. J Clin Oncol 10:180–199

Schendel DJ, Wank R, Dupont B (1979) Standardization of the human in vitro cell-mediated lympholysis technique. Tissue Antigens 13:112–120

Schendel DJ, Gansbacher B, Oberneder R, Kriegmair M, Hofstetter A, Riethmuller G, Segurado OG (1993) Tumor-specific lysis of human renal cell carcinomas by tumor-infiltrating lymphocytes. I. HLA-A2-restricted recognition of autologous and allogeneic tumor lines. J Immunol 151:4209–4220

Schendel DJ, Gansbacher B (1993) Tumor-specific lysis of human renal cell carcinomas by tumor-infiltrating lymphocytes: Modulation of recognition through retroviral transduction of tumor cells with interleukin 2 complementary DNA and exogenous a interferon treatment. Cancer Res 53:4020–4025

Schendel DJ, Oberneder R, Falk CS, Jantzer P, Kressenstein S, Maget B, Hofstetter A, Riethmüller G, Nößner E (1997) Cellular and molecular analyses of major histocompatability complex (MHC)-restricted and non-MHC-restricted effector cells recognizing renal cell carcinomas: problems and perspectives for immunotherapy. J Mol Med (in press)

Schwartz RH(1992) Costimulation of T lymphocytes: the role of CD28, CTLA-4, and B7/BB1 in interleukin-2 production and immunotherapy. Cell 71:1065–1068

Slingluff CL, Jr., Hunt DF, Engelhard VH (1994) Direct analysis of tumor-associated peptide antigens. Curr Opin Immunol 6:733–740

Tahara H, Zeh HJ, Storkus WJ, et al. (1994) Fibroblasts genetically engineered to secrete interleukin 12 can suppress tumor growth and induce antitumor immunity to a murine melanoma in vivo. Cancer Res 54:182–189

Van Pel A, Van der Bruggen P, Coulie PG, et al. (1995) Genes coding for tumor antigens recognized by cytolytic T lymphocytes. Immunol Rev 145:229–250

Veelken H, Jesuiter H, Mackensen A, et al. (1994) Primary fibroblasts from human adults as target cells for ex vivo transfection and gene therapy. Hum Gene Ther 5:1203–1210

6 Transplantation of Isolated Islets of Langerhans as Treatment of Diabetes Mellitus

K. Federlin, B.J. Hering, and R.G. Bretzel

6.1 Introduction

Transplantation of an immediately vascularized pancreatic allograft has been used for 30 years now as total endocrine replacement therapy for insulin-dependent diabetes mellitus. Especially if performed in conjunction with a kidney graft it dramatically increases the quality of life of the diabetic patient. The further therapeutic process rests on the hypothesis that the microangiopathy and macroangiopathy would prevent the development or halt the progression of defects of the eye, kidney, heart, peripheral arteries, and nervous system. However, pancreatic organ transplantation is a complicated operation which involves many risks and which is successful only in a few of the most experienced teams worldwide.

Historically, the first attempts of treating diabetes by transplantation of pancreatic tissue particles instead of the whole organ were reported already in 1892 by Minkowski. He described the reduction of glycosuria after subcutaneous implantation of small pieces of the autologous pancreas in pancreatectomized dogs. In 1967 Lacy and Kostianovsky described a method of isolating islets from the rat pancreas by tryptic digestion of the gland with collagenase. Soon after, Younoszai et al. in 1970 reported the first transplantation of such isolated islets from adult rat pancreas using outbred animals. In a few animals they observed a significant drop in urine glucose excretion and a reduction of the fasting blood glucose values on days 7 and 10. Later on, the islets were rejected. The first long-lasting successful results were reported by Ballinger and Lacy (1972) with inbred rats. A year later we presented the results of our first studies at the EASD Congress in Madrid (Federlin et al. 1973). Whereas the peritoneal cavity was used for the first implantation of isolated islets, Dr. Lacy's group demonstrated that islets injected into the portal vein resulted in lodging in the small branches of the vessel system in the liver with much better metabolic results (Kemp et al. 1973). Only 50% of the isolated islets used for intraperitoneal injection were necessary to normalize the blood glucose by intraportal transplantation. For the vast majority of experiments in smaller and larger animals as well as in clinical studies in humans this implantation site has been used up to the present.

6.2 Experimental Islet Transplantation in Rodents

Using inbred rats as donors and recipients of islets it could be demonstrated that the diabetic metabolism could be corrected for long periods of time (normoglycemia up to 2 years). The animals showed normoglycemia, euglycosuria, normalization of body weight and when transplanted early after diabetes induction by streptozotocin they were indistinguishable from normal animals (Fig. 1; Table 1).

Furthermore, it could be shown that after early islet transplantation diabetic rats did not develop typical diabetic late complications such as retinopathy, nephropathy, neuropathy, and other changes later on

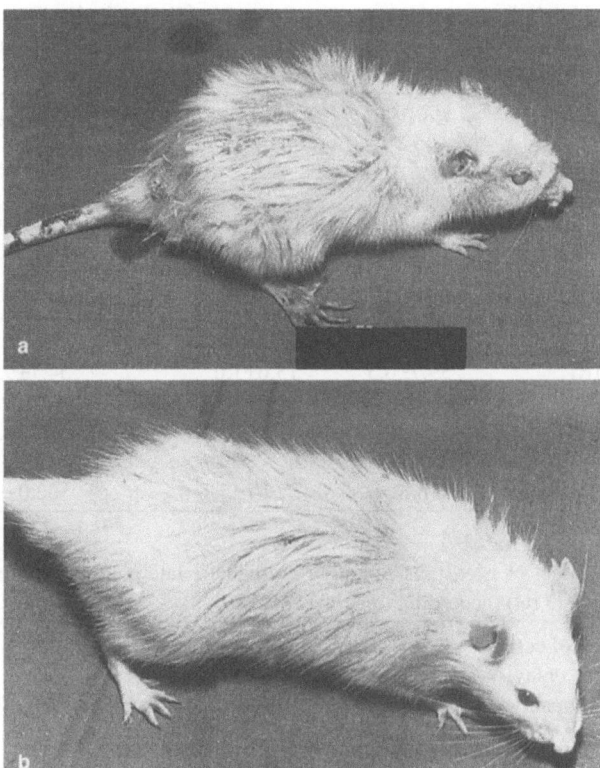

Fig. 1. a Diabetic rat 10 months after diabetes induction by streptozocin (65 mg/kg). **b** Animal of the same age treated by islet transplantation 4 weeks after diabetes induction. Normal shape, normal body weight

(Fig. 2; Table 2). This effect could be achieved using about 20%–25% of the islets of one rat pancreas (1000–1500) (Federlin et al. 1976; Bretzel et al. 1992).

Another approach for implanting isolated islets from the adult rat pancreas was the use of fetal pancreatic tissue, which offers the advantage of not having to isolate the islets because pancreatic tissue in the early period consists mainly of endocrine cells (Mandel 1992). The hope connected with the use of this kind of tissue was twofold: its prolifera-

Table 1. Metabolic parameters in rats at an age of 9 months in comparison to normal nontransplanted diabetic rats and with islet transplanted animals (according to Federlin 1993)

	Normals n=9	Diabetics no Transpl. n=20	Transplanetd n=16
Glycosuria (%)	0	2%	0 n=11 >2% n=5
Urine volume (ml/24h)	10.7±0.7	79.2±2.3	14.3±3.6
Fasting-BG (mg%)	78.2±11.1	337.3±20.9	76.1±6.3
Fasting-IRI (µU/ml)	39.7±6.3	13.6±1.6	54.7±9.5
k-value (i.v. GTT)	1.51±0.10	0.35±0.04	1.17±0.05
Body weight (gm)	416.8±4.3	229.7±5.7	362.0±9.8

Table 2. Effect of islet transplantation on diabetic late complication in the rat (Lewis-inbred rat) (according to Bretzel et al. 1992)

Prevention of cataract
Prevention, delay, or reversal of:
    retinopathy
    nephreopathy
    neuropathy
    cardiopathy
    skeletal disorders

tion capacity and reduced immunogenicity. The expectations in particular that fetal islet tissue would survive for a longer period in an allogeneic host were not fulfilled because, with maturation, immunogenicity increased to "normal."

The promising results of islet transplantation in diabetic inbred rats could, however, not be obtained in an allogeneic system. Intraportally injected islets between two rat strains with a major histocompatibility (MHC) barrier are rejected within a few days (Fig. 3). It became obvious that isolated islets are highly immunogenic. Several attempts have been made to reduce the immunogenicity of islets based on early studies of

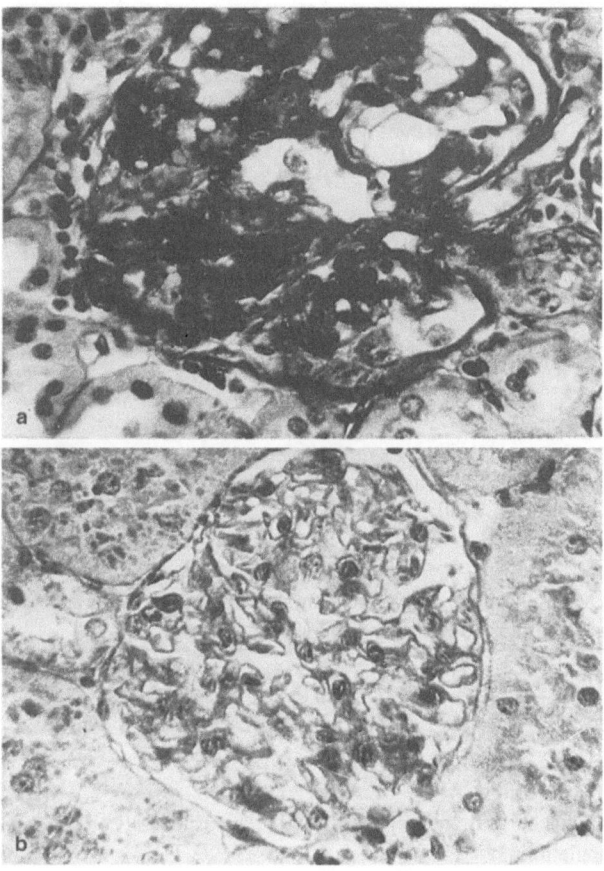

Fig. 2. a Glomerulosclerosis of the rat 12 months after diabetes induction. Enlarged mesangium with deposites of PAS-positive material, thickening of capillary walls. PAS staining, ×280. **b** Normal glomerulus in a primarily diabetic rat after islet transplantation 4 weeks after diabetes induction. Rat of the same age as in **a**. Loose arrangement of the glomerular capillaries, thin capillary membranes, but only scarce deposition of PAS-positive material. PAS staining, ×280

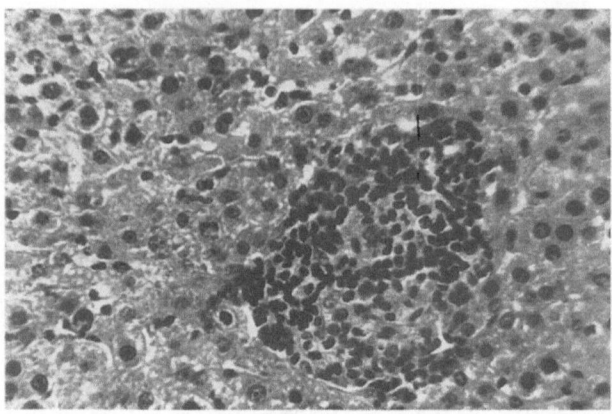

Fig. 3. Rejection of an allogeneic islet 5 days after intraportal transplantation in the rat. Masive cell infiltration of the islet, only few surviving β-cells. PAP staining for insulin, ×360

Table 3. Methods known for in-vitro immunoalteration of adult pancreatic islets of the rat (according to Bretzel et al. 1986)

γ-irradiation
UV-light
Immunosuppressive drugs
Gangliosides
Antineoplastic drugs
MoAb against class-II antigen bearing cells, dendritic cells or lymphocyte
    subpopulations
Cyropreservation
Culture 22°C, 24°C, 37°C, Hyperbaric/high oxygen, megaislets, pseudoislets

Lafferty and his group (1976), who showed that thyroid allografts in mice had a longer survival after a culture period at 22°C. The theory is that low temperature destroys or masks class II antigens on the surface of immunocompetent cells such as dendritic cells, macrophages, lymphocytes, endothelial cells, and others. Rat islets treated accordingly survived up to 200 days in an allogeneic host without any immunosuppressive treatment (Lacy et al. 1979). Furthermore, some other pretreatment schedules reduced rat islet immunogenicity with increased survival time in allogeneic hosts (Table 3). Unfortunately, up to now immunoalteration of islets of larger animals or humans has not been successful as of yet. Nevertheless, the principle of changing cell surface properties of islets of Langerhans even by single or by a combined immunoalteration methods still offers some hope for the future.

There have been also several attempts to transplant islets into so-called immunologically privileged sites where presentation and recognition of allogeneic MHC antigen does not take place because of the lack of a lymphatic drainage system (cheek pouch of the hamster, anterior chamber of the eye, testis, and brain). For islets, however, those sites did not offer any real advantage.

Very recently rat islet allograft was transplanted into the thymus of recipients treated with a single injection of antilymphocytic serum (Posselt et al. 1990). Those islet allografts survived indefinitely. In this model a state of donor-specific unresponsiveness was achieved which permitted survival of a second donor strain islet allograft transplanted to an extrathymic site. The authors speculate that maturation of precursors in a thymic microenvironment that is harboring an alloantigen may induce the selective unresponsiveness. Although this elegant technique raised much hope for further studies with allogeneic islet grafts, up to now this rodent model has not worked as well in larger animals.

6.3 Islet Transplantation in Larger Animals

Autotransplantation and allotransplantation have been also performed in pigs, dogs, and monkeys. When isolated islets of pigs were initially transplanted in the peritoneal cavity of totally pancreatectomized recipients, only a minimal effect on glucose metabolism was observed (Sutherland et al. 1974). Using more purified preparations of autologous islets transplanted into the spleen or intraportally (own observations, unpublished), normoglycemia could be observed for several weeks. Up to now, however, a more widespread success of islet transplantation in pigs has not yet been achieved for several reasons (difficulties in mass isolation of highly purified islets and problems with long-term keeping of pancreatectomized animals). In recent years auto- and allotransplantation in pigs have been extensively studied by the group of Professor

Hopt (Tuebingen/Rostock, Germany). Special emphasis was placed on the rejection process of allogeneic islets after intraportal transplantation as comparable to islet transplantation in humans. Interestingly, even with the strongest immunosuppressive regimen it was not possible to definitely prolong islet function (Mellert et al. 1990).

Islet transplantation studies in dogs were more promising. The most reliable model for accurate assessment of islet graft function is insulin-dependent diabetes induced by total pancreatectomy, which, however, requires extensive experience on the part of the surgeon. The largest series of studies were performed in Miami (Alejandro et al. 1985) and in Edmonton (Warnock et al. 1988). As the summary of these investigations it was found that about 5000 islet equivalents per kilogram of body weight are necessary to consistently reverse diabetes after autotransplantation into pancreatectomized recipients with a body weight of 20–25 kg. The Miami group was able to demonstrate that long-term survival of allogeneic islets could also be observed when cultured islets plus cyclosporine A was used. Furthermore, the Edmonton group studied in detail the use of cryopreserved purified dog islets in comparison to freshly prepared islets. It could be shown that an increase of about 20% of the mass of islets for cryopreserved material was necessary to reach the same effect as with fresh islets. However, it was possible to maintain euglycemia for several months by highly purified cryopreserved autologous islets (Evans et al. 1990). Interestingly, the islet mass required in the canine model of pancreatectomy-induced diabetes was very similar to the islet mass which was necessary to induce euglcemia in human diabetics when diabetes was due to pancreatectomy (Tzakis et al. 1990). In those patients islet allotransplantation was performed via the portal vein into the transplanted liver from the same donor, which means an organ which has not been damaged by long-lasting diabetes. The situation in type 1 diabetes may be different because the possible long-term effect of hyperglycemia on the vascular system of the liver and because of the autoimmune thread for healthy islets.

In contrast to the experimental results in rodents, the long-term effect of islet transplantation in diabetic dogs and pigs is still lacking in larger animals. Interestingly, the intrasplenic canine autograft showed a much longer function than islets transplanted intraportally (4 years vs 2 years) (Warnock et al. 1992a). The reason for this is not clear. It has been suggested that islets within the liver could be exhausted by the high portal glucose concentration. Unfortunately, there is no autoimmune diabetic model available in dogs or pigs as in the BB rat or the NOD mouse which could serve as a parallel to the human situation of type 1 diabetes.

Transplantation studies with islets in primates were also limited regarding their number. In pancreatectomized baboons persistent hyperglycemia was reported after autologous islet transplantation in the renal subcapsular region (Du Toit et al. 1984). Another group used the pancreatectomized cynomolgus monkey and reversed diabetes by intrasplenic or intraportal transplantation of the purified autologous islets (Sutton et al. 1987). Recently, the monkey has become more attractive in islet transplantation research as a model for xenotransplantation (porcine islets to monkeys). Whether those ongoing studies in several laboratories will be successful has to be awaited.

6.4 Islet Transplantation in Humans

Clinically, islet transplantation in humans is performed routinely by catheterization of a branch of the portal system transcutaneously under X-ray control. When the catheter is placed in the lumen of the portal vessel the islets are infused within 25–30 min (500 000 islets dissolved in about 100 ml of Ringer's solution). Intermittently, the portal pressure is controlled and the infusion slowed down when portal pressure increases for a short period. The procedure is rather safe. In one case portal vein thrombosis was observed following transplantation of impure islet tissue and another death occurred after bleeding from an hepatic arteriola branch in a patient with four previous cardiac infarctions. Although the bleeding was controlled, the patient died due to cardiac insufficiency (own observation, in press).

Islet transplantation in humans can be divided into three categories:

1. Islet autotransplantation
2. Islet allotransplantation in pancreatectomized patients
3. Islet allotransplantation in type 1 diabetics

6.4.1 Islet Autotransplantation

Recurring pancreatitis, especially due to alcoholism, may lead to severe
continously or recurring pains which are nearly untractable. The only
treatment providing definite pain relief is pancreatectomy, which, however, has the consequence of inducing a type of diabetes which is
difficult to manage. The lack of the counterregulatory hormone glucagon, in particular, is a main reason for the usually sudden changes
from hyperglycemia to hypoglycemic episodes during the insulin treatment of those patients. The isolation of islets from the completely
removed gland and transplantation as a free graft may help to avoid
diabetes as well as diabetic complications in the future. In the last two
decades several attempts have been made to treat patients with chroni-
cally recurring pancreatitis by islet autotransplantation. In 1978 the first
patient at Giessen University received his own islets in the form of
dispersed pancreatic fragments which were transplanted into the portal
vein (Dobroschke et al. 1978). Although the operation was safe, it did
result only in some metabolic effect: The patient did not become insulin
independent but even after nearly 20 years his insulin requirement is still
low (10–12 U/day). In the meantime the technique of isolation has been
refined, especially by the automated method of Ricordi et al. (1988),
which resulted in a much better outcome in most of the patients. Though
the number of cases is still low, the Minneapolis group published remarkable data (Pyzdrowski et al. 1992). They reported on islet autografts in 24 patients, seven of which became insulin independent for
more than 1 year. According to the International Islet Transplant Registry, the longest follow-up of a patient with an islet autograft and insulin
independence is now more than 13 years (Table 4). In addition, longer-
lasting, normal, glycosylated hemoglobin A1c levels were reported.
There was no complication in the islet recipients such as disseminated
intravascular coagulation, significant portal hypertension, or hepatic
dysfunction. Needle biopsy revealed clusters of islet staining strongly
for insulin and glucagon within the portal triads. This result is especially
encouraging for future islet transplantation in man because it serves as a
prove for a long-term normal glucose homeoestasis.

Table 4. Islet autografts: insulin independence after retransplanation of more than 300 000 islets (from Hering et al. 1995)

No. of cases from 1974 through Dec. 31, 1995	163
Insulin independence	(77.4%) 24/31[a]
Insulin independence at ≥ 1 yr	(66.6%) 12/18[a]
Longest follow-up of insulin independence	>13 yrs

[a]Only well documented patients

Summary of Adult Islet Autografts According to Institution and Year through December 31, 1995

Institution (Islet Transplantation/ Islet isolation)	Year of Transplantation							
	>90	90	91	92	93	94	95	Σ
Minneapolis	23	3	1	2	4	12	8	53
Genova	24	–	–	–	–	–	–	24
Berlin (Charité)	12	–	–	–	–	–	–	12
Geneva	–	–	–	2	2	1	4	9
Leicester	–	–	–	–	–	1	8	9
Baltimore	8[a]	–	–	–	–	–	–	8
Detroit	8	–	–	–	–	–	–	8
Los Angeles (UCLA-VA)	5			–	–	1		6
Boston	5[a]	–	–	–	–	–	–	5
Loma Linda	5	–	–	–	–	–	–	5
Nagasaki	5	–	–	–	–	–	–	5
Giessen	4	–	–	–	–	–	–	4
Pittsburgh	–	–	2	1	–	–	1	4
Helsinki	2	–	–	–	–	–	–	2
Philadelphia	2[a]	–	–	–	–	–	–	2
St. Louis	2	–	–	–	–	–	–	2
Charlotte	–	–	–	–	–	–	1	1
Gent/Giessen	–	–	–	–	1	–	–	1
Leningrade	1[a]	–	–	–	–	–	–	1
New York	1[a]	–	–	–	–	–	–	1
Paris	1[a]	–	–	–	–	–	–	1
Σ	108	3	3	5	7	14	23	163

[a]Year published

6.4.2 Islet Allotransplantation in Pancreatectomized Patients

Dr. Starzl's group in Pittsburgh used islet allotransplantation in a group of patients in which upper abdominal exenterations for tumors of stomach, liver, pancreas, and the total region was performed. They received a cadaveric autotopic liver allograft and the islets of the same donor pancreas were injected via the portal vein. The patients received an average of implanted 392 000 islets and five of seven patients did not require insulin for a period of more than 1 year, the longest period reaching 56 months (Table 5). Some patients died of tumor recurrence but did not require insulin at that time. Another type of transplantation in primarily nondiabetic patients is hemochromatosis, which led to a simultaneous islet–liver transplant. Two of those patients were treated in Paris. The islets were embolized to an epiploic flap by using an arteriolar branch of the right gastroepiploic vessels. One patient showed normal baseline stimulated C-peptide levels after transplant and became insulin independent in the seventh postoperative month and has been off insulin for several years now. Patients of this category may have a better outcome of islet allotransplantation because the islets are transplantated into a nondiabetic liver and they are not threatened by diabetes autoimmunity of the recipient organism.

6.4.3 Islet Allotransplantation in Type 1 Diabetics

A few years after the first successful attempts to treat streptozocin-induced diabetes in rats by transplantation of isolated islets, the Minneapolis group initiated clinical islet allotransplantation in diabetic patients. In 1977 the first attempts of intraperitoneal and intraportal transplantation of dispersed human pancreas from adult or infant donors in seven patients with type 1 diabetes were reported (Najarian et al. 1977). No patient became insulin independent but in four patients there were brief intervals of reduced insulin requirements. The main message of this pioneer work was that human islets also started insulin secretion in a heterotopic environment (liver or peritoneal cavity) and that the transplantation procedure itself was safe. Furthermore, no rejection crisis of a previously transplanted kidney occurred. From a rough calculation one could estimate that the number of islets transplanted in this first series ranged between 15 000 and 150 000. So at this time it could be argued that the number of the islet mass was too low to reach insulin independence. In the following years the number of adult islet allografts was low but showed an increase in 1989/1990 after the development of a semi-automatic islet isolation method and a new purification technique with the Cobe cell separator by Ricordi (Ricordi et al. 1988). This new technology has permitted a much higher yield of purified and viable islets – up to 800 000 from one pancreas (own results, unpublished).

During this first era of islet transplantation in humans there was only one patient in whom hope was raised that insulin independence could be achieved by this type of treatment. Largiader et al. (1979) reported the case of a long-term functioning human pancreatic allotransplant. The female patient received a renal graft and pancreatic islet transplants from an immature donor. The islets, seemingly nonfunctioning at first, survived three reversal rejection episodes during the first 6 months after operation and developed endocrine sufficiency within the following 2 months. The patient received the islets on 29 August 1978 and died suddenly from intestinal hemorrhage on 29 April 1980. From June 1979 until her unexpected death she was insulin independent and the rejection therapy needed only azathioprine (50 mg/day) and prednisone (5 mg/day). More than 10 years passed until the St. Louis group (Scharp et al. 1990) observed insulin independence again in a type 1 diabetic patient with an established kidney transplant. He received 800 000 islets from one and a half pancreata from cadavers. Immunosuppression was performed with antilymphoblast globulin for 7 days and maintained with cyclosporine. Although the patient rejected the transplant after 14 days this was the first demonstration of the ability of purified human islet transplant given intraportally as an allograft to achieve exogenous insulin independence (in this case there was no measurement of C peptide to clearly document islet function). More promising was the long-term effect of an islet transplant which was reported by the Edmonton group (Warnock et al. 1992b). At this time the patient was insulin independent for 1 year starting 69 days after intraportal grafting of freshly isolated and cryopreserved islets. This remarkable patient remained insulin free for another year until the exogenous insulin treatment was reinstalled. This case demonstrated for the first time definitely longer-lasting function of intraportal transplanted islets and, in addition, the practicability of cryopreserved islets.

Table 5. Insulin independence (1 week) after adult islet transplantation into patients with pancreatectomy-induced diabetes mellitus: summary of cases through 31 December 1994 (from Hering et al. 1996)

Institution	Year of Tx	No. of Donors	IEQ/kg	Recipient Category	Site of Tx	Induction Immunosuppression	Period of Insulin Independence (Months Post-Tx)		
Pittsburgh	1990	1	11,850	SIL	liver	FK 506 + P (single bolus)	2	–	58[a]
Pittsburgh	1990	1	6,568	SIL	liver	FK 506 + P (single bolus)	2	–	4
Pittsburgh	1990	2	8,028	SIL	liver	FK 506 + P (single bolus)	2	–	15
Pittsburgh	1990	2	4,526	SIL	liver	FK 506 + P (single bolus)	5	–	15
Pittsburgh	1990	1	3,519	SIL	liver	FK 506 + P (single bolus)	3	–	101
Pittsburgh	1990	2	15,447	SIL	liver	FK 506 + P (single bolus)	2	–	20[a]
Pittsburgh	1990	2	5,627	SIL	liver	FK 506 + P (single bolus)	1	–	15[a]
Milan	1992	2	11,526	SIL	liver	ALG (+P-C-A)	2	–	14
							21	–>	25[b]
Giessen/ Wuerzburg	1993	2	9,571	SIL	liver	ALG + Anti-IL-2 (+P-C)	4	–	6

[a]Died off insulin, [b]as of Dec. 31, 1994.

P, Prednisone; IEQ, Islet Equivalents; C, Cyclosporine A; SIL, Simultaneous Islet Liver; A, Azathioprine.

Since the International Islet Transplant Registry in cooperation with the International Pancreas Transplant Registry of Dr. D.E.R. Sutherland, Minneapolis, was installed at Giessen University in 1989, seven newsletters have appeared reporting the results which were obtained worldwide in islet transplantation centers. A summary of adult islet allografts (and one xenograft) according to institution and year through 31 December 1995 is given in Table 5. At this time 306 cases had been reported as having been done by 35 institutions. As shown in Fig. 4 the number of adult islet allografts and institutions reporting to the Registry has increased in recent years. In addition, it is assumed that today over 2000 fetal islet allografts and xenografts have been performed worldwide (mainly in Russia and China). In the meantime, 187 fetal islet allografts have also been reported to the Registry. So far, insulin independence has not been achieved in a pretransplant C-peptide-negative, type 1 diabetic recipient. The pretransplant C-peptide levels in the three patients who became insulin independent after fetal islet tissue transplantation were not determined. However, recent reports demonstrate

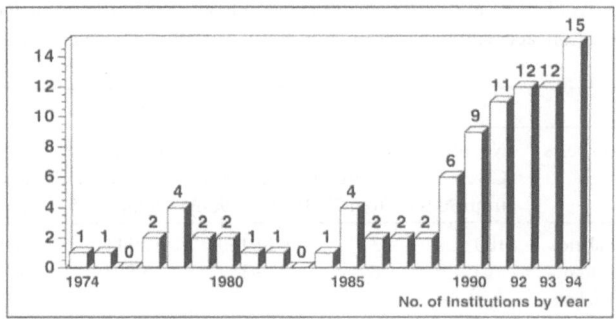

Fig. 4. Number of institutions reporting adult islet allografts by year from 1974–31 December 1994. From Hering et al. (1996)

basal C-peptide levels exceeding 0.5 ng/ml in eight recipients without residual C-peptide secretion pretransplant lasting for periods of 2–48 months. More details have been described in Newsletter No. 6 of the International Islet Transplant Registry (Hering et al. 1995).

The total number of diabetic patients reported to be insulin independent for ≥1, ≥3, ≥6, ≥12, ≥24, ≥36, and ≥48 months after adult islet allotransplantation through 31 December 1995 is 39, 36, 26, 24, 16, 10, 4, and 1, respectively. In an analysis by era, the percentages of pretransplant C-peptide negative type 1 diabetic recipients of adult islet allografts who showed basal C-peptide levels ≥1 ng/ml at ≥1 month posttransplant and who became insulin independent for more than 1 week in the 1985–1989 era (n=28) were 25% and 7%, in the 1990–1994 era (n=131) were 57% (n=75) and 20% (n=26), respectively (p=0.003, p=0.169) (Hering et al. 1996).

As induction immunosuppression ALG/ATG or OKT3, steroids, cyclosporine or azathioprine were used and in recent years FK506, desoxyspergualine and mycophenolate mofetil (MMF) have been used. As the maintenance immunosuppression steroids, cyclosporine, and azathioprine or, again, FK506, MMF have been reported.

More recent data on insulin-independent type 1 diabetic recipients are shown in Tables 6–9. While a total of 27 patients from 270 became insulin independent for sometime and showed independence for more than 1 year and the long-risk observation period for a patient who could

Table 6. Adult islet allografts: insulin dependence following adult islet allo-transplantation according to recipient's type of diabetes (from Hering et al. 1996)

Type of Diabetes	No. of Cases from 90-95	No. of Institutions from 90-95	Insulinindependence		
			≥1 wk all Cases from 90-95	≥1 wk after 1:1 Tx from 90-95	at ≥1 yr all Cases from 90-95
Type 1 Diabetes	180	23	24/180	12/111	11/180
Pancreatectomy-Induced Diabetes	15	6	9/15	3/8	6/15
"Insulin Requiring" Diabetes (type 2, cystic fibrosis, hemochromatosis)	12	7	3/12	3/9	1/12

discontinue taking exogenous insulin was 37 months. So the principal feasibility of islet allotransplantation in autoimmune diabetes leading to insulin independence has been proven by various institutions.

However, even partial success of islet transplantation which is given when a patient was C-peptide prior to transplantation can mobilize at least some insulin, as proven by constantly present C-peptide levels. The chance to reach normal glycohemoglobin levels and to avoid hypoglycemia also represents a big step forward for this type of patients, without having reached the ultimate goal of insulin independence. As shown in Fig. 5, 12 months after transplant basal C-peptide levels could be demonstrated in 92 patients (28%) who were C-peptide negative before transplantation.

The poor outcome of treatment with islet transplantation in type 1 diabetics as measured by insulin independence raises many questions regarding the definite causes. The observation that surgical diabetes (patients with pancreatectomy) is easier to cure than type 1 diabetes underlines that alloimmunity itself may not constitute the primary obstacle. Two main facts emphasize the key role of autoimmunity: the recurrent diabetes in three patients with a 17- to 36-year history of type 1 diabetes 4–12 weeks after transplantation of a segmental pancreas from HLA-identical twins due to isletitis with selective destruction of islet β-cells (Sibley et al. 1985) and, secondly, the onset of diabetes in a

Table 7. Summary of adult islet allografts (and one xenograft) according to insitution and year through 31 December 1995 (from Hering et al. 1996)

Institution (islet Transplantation/ Islet isolation)	Year of Transplantation							
	<90	90	91	92	93	94	95	Σ
Minneapolis	24	1	4	5	5	2	10	51
Pittsburgh	17	6	4	3	4	3		37
St. Louis	13	3	3	2	4	2	–	27
Miami	5	4	2	1	1	3	9	25
Giessen	–	–	–	1	5	5	12	23
Milan	1	4	3	3	4	4	4	23
Genova	13	–	–	–	–	–	–	13
Berlin (Charité)	8	–	–	–	–	–	–	8
Zurich	8	–	–	–	–	–	–	8
Detroit	7	–	–	–	–	–	–	7
Edmonton	2	2	–	1	–	1	1	7
Oxford	–	–	1	1	1	1	3	7
Paris	3	3	1	–	–	–	–	7
Perugia	3	1	1	–	–	2	–	7
Madrid	–	–	–	2	1	1	2	6
Brussels	–	–	·	·	–	1	4	5
Odense/Milano	–	–	–	–	–	–	5	5
San Francisco/LA (UCLA-VA)	–	–	–	–	1	1	3	5
London (Ontario)/St. Louis	–	2	1	1	–	–	–	4
Leicester	–	–	2	1	–	–	–	3
Los Angeles (UCLA-VA)	–	–	–	3	–	–	–	3
Padova/Verona	2	–	–	–	1	–	–	3
Los Angeles (St. Vincent)	–	–	–	–	2	1	–	3
Charlestown	–	–	2	–	–	–	–	2
Hanover	2	–	–	–	–	·	–	2
Newcastle-upon-Tyne	2	–	–	–	–	·	–	2
Berlin (Moabit)	1	–	–	–	–	–	–	1
Bristola	1	–	–	–	–	–	–	1
Geneva	–	–	–	–	–	1	–	1
Heidelberg/Giessen	1	–	–	–	–	–	–	1
Homburg	–	–	–	–	1	–	–	1
Innsbruck/Milano	–	–	–	–	–	–	2	2
Leiden	1	–	–	–	–	–	–	1
London (Ontario)	–	–	–	–		1	–	1
New York	1	–	–	–	–	–	–	1
Omaha	–	–	–	–	–	1	–	1
Petah Tiqwah	1	–	–	–	–	–	–	1
Würzburg/Giessen	–	–	–	–	1	–	–	1
Σ	99	37	26	25	30	31	58	306

Table 8. Type 1 diabetic recipients of adult islet allografts: summary of insulin-independent cases ($n=27$) through 31 December 1995, part 1 (from Hering et al. 1996)

Institution	Year of TX	Previous Pancreas/ Islet Tx	Pre-Tx C-Pept. (ng/ml)	No. of Donors Fresh	Cyro	IEQ/kg	Purity (%)	Period of Insulin Independence Days post Tx
Zurich	1978	–	NA	1	–	3,846[a]	5	290 – 590
Paris[b]	1988	–	S 0.03	1	–	2,143[a]	80	214 – 1,495
St. Louis	1989	P 83	S 0.06	2	–	12,661	95	10 – 25
St. Louis	1990	I 89	S 0.54	1	+ 2	14,733	98	33 – 341
St. Louis	1993	I 89, I 90	S 1.26	2	+ 5	23,733	92	92 -> 662
St. Louis	1993	–	S 0.24	3	+ 2	27,370	87	274 – 355
Edmonton	1990	–	S 0.00	1	+ 4	9,698	70	69 – 821
Edmonton	1992	–	S 0.00	1	+ 5	9,867	58	155 – 166 + 37 – 992
Milan	1990	–	S 0.00	1	–	10,767	95	120 – 330
Milan	1990	P 88	B 0.15	2	–	8,607	78	60 – 1,178
Milan	1991	P 87	B 0.00	1	+ 2	15,241	80	210 – 360 + 480 – 635
Milan	1992	–	B 0.00	2	–	11,566	80	150 -> 1,122
Milan[c]	1994	P 91, P 92	B 0.15	1	+ 2	28,995	55	42 – 65 + 92 – 133
Milan[c]	1995	P85	B 0.00	1	–	9,600	50	56 – 100
Miami	1990	–	S 0.03	3	–	18,700	55	42 – 78
Miami	1990	–	S 0.03	3	–	18,891	50	87 – 125
Miami	1995	–	B 0.03	1	–	15,691	85	49 – 69
Minneapolis	1992	–	S 0.09	1	–	7,882	5	326 – 1,241
Minneapolis	1992	–	S 0.39	1	–	13,319	5	123 – 321
Minneapolis	1995	–		1	–	8,998	5	43 -> 144
Giessen	1992	–	S 0.10	1	–	6,158	92	400 -> 1,243
Giessen	1995	–	S 0.00	1	–	7,789	95	12 – 25
Giessen	1995	–	S 0.00	1	–	11,148	90	312 -> 495
Giessen	1995	–	S 0.06	1	–	7,760	90	363 -> 469
Padua/Verona[d]	1993	–	B 0.62	1	–	4,392	60	35 -> 532
Pittsburgh	1994	–	s 0.00	1	–	2,143	80	118 – 723
Odense/Milan[d]	1995	–	s 0.00	2	–	9,360	80	79 -> 138

P, Pancreas; I, Islet; S, Stimulated; B, Basal.
[a]Islets/kg ; IEQ: Islet Equivalents (150 µm islets).
[b]Cholangiocarcinoma, hemochromatosis, and type 1 diabetes.
[c]Previous transplants and follow-up in Nantes.
[d]Islet Transplant/Islet Isolation Institution.

Table 9. Type 1 diabetic recipients of adult islet allografts: summary of insulin-independent cases (n=27) through 31 December 1995, part 2 (from Hering et al. 1996)

Institution[a]	Year of Tx	Site of Tx	Recipient Category	# of Shared HLA-AA AB	DR	Immunosuppression Induction		Maintenance
Zurich	1978	spleen	SIK	1	0	ALG	+S+A+CPM	S+A
Paris	1988	epiploic flap	SIL	1	1	ALG	+S+C+A	S+C+A
St. Louis	1989	liver	IAK	1/3	2/1	ALG	+S(SD)+C+A	C+A
St. Louis	1990	liver	IAK	1/2/2	1/i/O	ALG	+S+A	S+A
St. Louis	1993	liver	SIK	1/0/0/0/1/0	1/0/1/0/1/0/1	OKT3	+S+C+A	S+C+A
St. Louis	1993	liver	SIK	1/l/2/l/1	1/l/l/2/1	ATG	+S+C+A	S+C+A
Edmonton	1990	liver	SIK	3/1/0/1/0	0/0/0/0/0	ALG	+S+A	S+C+A
Edmonton	1992	liver	SIK	3/1/0/0/1/0	1/1/1/0/0/1	ALG	+S+A	S+C+A
Milan	1990	liver	IAK	1	0	ALG	+S+C+A	S+C+A
Milan	1990	liver	IAK	1/2	1/0	ALG	+S+C+A	S+C+A
Milan	1991	liver	SIK	NA	NA	ALG	+S+C+A	S+C+A
Milan	1992	liver	IAK	NA	NA	ALG	+S+C+A	S+C+A
Milan	1994	liver	IAK	0/l/2	NA	ATG	+S+C+A	C+A
Milan	1995	liver	IAK	0	NA	ATG	+S+C+A	S+C+A
Miami	1990	liver	IAK	0/2/0	1/1/0	OKT3	+S+C+A	S+C+A
Miami	1990	liver	IAK	0/0/0	0/1/0	OKT3	+S+C+A	S+C+A
Miami	1995	liver	SIK	0	0	ATG	+S+T+MMF	S+T+MMF
Minneapolis	1992	liver	SIK	I	1	ALG	+S+C+A+D	S+C+A
Minneapolis	1992	liver	SIK	2	0	ALG	+S+C+A+D	S+C+A
Minneapolis	1995	liver	IFPP	2	1	None	+S+C+MMF	S+C+MMF
Giessen	1992	liver	IAK	2	1	ATG	+S+C	S+C
Giessen	1995	liver	ITA	1	0	anti-CD4	+S(SD)+C	–
Giessen	1995	liver	SIK	1	1	ATG	+S+C+A	S+C+A
Giessen	1995	liver	SIK	2	1	ATG	+S+C+A	S+C+A
Padua/Verona	1993	liver	SIL	NA	NA	None	S+C	S+C+A
Pittsburgh	1994	liver	SIK	3	2	None	S+T	S+C+A
Odense/Milan	1995	liver	SIK	2/NA	1/NA	ATG	+C	C

SIK, Simult. Islet Kidney; SIL, Simult. Islet Liver; IAK, Islet After Kidney; IFPP, Islets From Previous Pancreas; S, Steroids; SD, Single Dose; C, Cyclosporin A; A, Azathioprine; MMF, Mycophenolate Mofetil; T, Tacrolimus; CPM, Cyclophosphamide; D, Deoxyspergualin.

[a]Islet Transplant/Islet Isolation Institution.

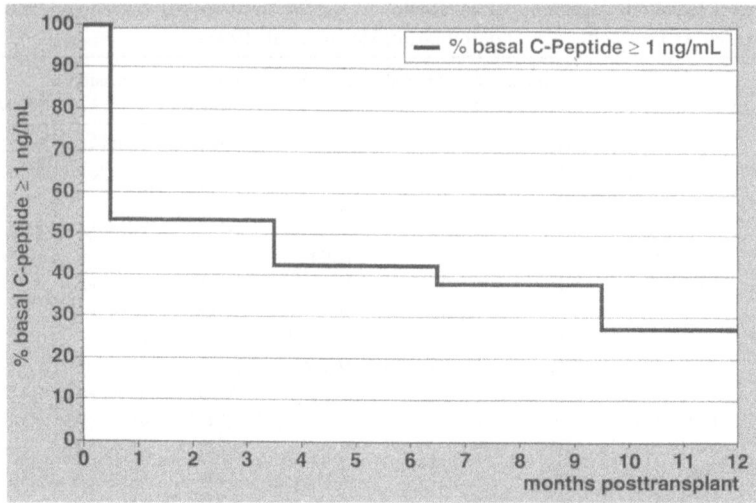

Fig. 5. One-year islet allograft survival in 92 pretransplant C-peptide-negative insulin-dependent diabetes mellitus recipients. Basal C-peptide ≥1 ng/ml, indicating insulin secretion after islet transplantation in primarily C-peptide negative diabetic recipients. From B.J. Hering, ITR (unpublished, in preparation)

woman 4 years after the transplantation of bone marrow from her HLA-identical brother with an 8-year history of type 1 diabetes (Lampeter et al. 1993). Thus, autoimmunity in association with alloimmunity constitutes the central problem in islet transplantation in type 1 diabetics.

6.5 Conclusion

Islet transplantation will only achieve significant levels of success in the not too distant future if techniques for the early detection of islet rejection episodes can be developed and if immunosuppressive regimens can be elaborated which will protect islet allografts likewise from recurrence of autoimmunity and classic T-cell mediated rejection but which will not imped insulin secretion and insulin action significantly.

Nevertheless, for the time being diabetic patients expecting a kidney graft are candidates which may have a definite advantage from a simultaneous islet graft from the pancreas of the same organ donor. In our studies it has been clearly shown that in comparison with patients receiving a kidney alone recipients with kidney plus islet grafts had much better glycohemoglobin levels after 1 year and the frequency of hypoglycemic attacks dropped dramatically. No patient of this category has become insulin independent yet but the partial success is clinically most relevant. Further hopes are connected with the use of new immunosuppressive agents such as Mycophenolat Mofetil, which has been found to be successful in organ transplantation of kidney and heart. In conclusion, islet transplantation is a safe procedure and, in principle, a logical treatment in type 1 diabetes. It provides a new chance for a minor number of patients with type 1 diabetes for allografts and would be a hope for an even larger number if xenotransplantation could be introduced.

References

Alejandro R, Cutfield R, Shienvold FL, Latif Z, Mintz DH (1985) Successful long-term survival of pancreatic islets allografts in spontaneous or pancreatectomy-induced diabetes in dogs. Diabetes 34:825–828

Ballinger WF, Lacy PE (1972) Transplantation of intact pancreatic islets in rats. Surgery 72:175–186

Bretzel RG, Blum BE, Höll E, Hering BJ, Federlin K (1986) Rat islet allograft survival following different immunomodulative and immunosuppressive treatment. Exc Med ICS 717:181

Bretzel RG, Hering BJ, Stroedter D, Zekorn T, Federlin KF (1992) Experimental islet transplantation in small animals. In: Ricordi C (ed) Pancreatic islet cell transplantation. Landes, Georgetown, pp 249–260

Dobroschke J, Schwemmle K, Langhoff G, Laube H, Bretzel RG, Federlin K (1978) Autotransplantation von Langerhans'schen Inseln nach totaler Duodenopankreatektomie bei einem Patienten mit chronischer Pankreatitis. Dtsch Med Wochenschr 103:1905–1910

Du Toit DF, Heydenreych JJ, Smit B, Louw G, Zuurmond T, Laker L, Els D, Weideman A, Wolfe Coote S, van der Merwe EA (1984) Experimental vascularised segmental pancreatic islet transplantation in the baboon. World J Surg 8:236–243

Evans MG, Warnock GL, Kneteman NM, Rajotte RV (1990) Reversal of diabetes by transplantation of pure cryopreserved islets. Transplantation 50:202–206

Federlin K (1993) Islet transplantation. The connection of experiment and clinic exemplified by the transplantation of islets of Langerhans. Berthold Memorial Lecture. Exp Clin Endocrinol 101:334–345

Federlin K, Helmke K, Slijepcevicz KM, Pfeiffer EF (1973) Transplantation of isolated islets of Langerhans in pancreatectomized rats. Diabetologia 9:66

Federlin K, Bretzel RG, Schmidtchen U (1976) Islet transplantation in experimental diabetes of the rat. V. Regression of glomerular lesions in diabetic rats after intraportal transplantation of isogeneic islets. Preliminary results. Horm Metab Res 8:404–406

Hering BJ, Schultz AO, Geier C, Bretzel RG, Federlin K (1995) Newslett Int Islet Transplant Registry 5 (6/1):

Hering BJ, Brendel MD, Schultz AO, Schultz B, Bretzel RG (1996) Newslett Int Islet Transplant Registry 6 (7):

Kemp CB, Knight MT, Scharp DW, Lacy PE, Ballinger WF (1973) Effect of transplantation site on the results of pancreatic islet isografts in diabetic rats. Diabetologia 9:486–491

Lacy PE, Davie JM, Finke EH (1979): Prolongation of islet allograft survival following in vitro culture (24°C) and a single injection of ALS. Science 204:312–313

Lacy PE, Kostianovsky M (1967) Method for the isolation of intact islets of Langerhans from the rat pancreas. Diabetes 16: 35-40

Lafferty KJ, Bootes A, Dart G, Talmage DW (1976) Effect of organ culture on the survival of thyroid allografts in mice. Transplantation 22:138–149

Lampeter EF, Homberg M, Quabeck K, Schäfer UW, Wernet, P, Bertrams J, Grosse-Wilde H, Gried FA, Kolb H (1993) Transfer of insulin dependent diabetes between HLA-identical siblings by bone-marrow transplantation. Lancet 341:1243–1244

Largiader F, Kolb E, Binswanger U, Billig R (1979) Erfolgreiche Pankreas-Inselallo-Transplantation. Schweiz Med Wochenschr 109:1733–1736

Mandel TE (1992) Immunobiology of the fetal and neonatal islet. In: Ricordi C (ed) Pancreatic islet cell transplantation. Landes, Georgetown, pp 26–36

Mellert J, Hering BJ, Brandhorst H, Klitscher D, Hufnagel B, Bretzel RG, Hopt UT, Federlin K (1990) Experience with islet allografts in immunosuppressed pigs. Horm Metab Res 25:187–189

Minkowski O (1892) Weitere Mitteilungen über den Diabetes mellitus nach Exstirpation des Pankreas. Berl Klin Wochenschr 29:90–94

Najarian JS, Sutherland DER, Matas AJ, Steffes MW, Simmons RL, Goetz FC (1977) Human islet transplantation; a preliminary report. Transplant Proc 9:233–236

Posselt AM, Barker CF, Tomaszewski JE, Markmann JF, Choti MA, Naji A (1990) Induction of donor-specific unresponsiveness by intrathymic islet transplantation. Science 249:1293–1295

Pyzdrowski KL, Kendall, Halter JB, Nakhleh RE, Sutherland DER, Robertson RP (1992) Preserved insulin secretion and insulin independence in recipients of islet autografts. N Engl J Med 327:220–226

Ricordi C, Lacy PE, Finke EH, Olack BJ, Scharp DW (1988) Automated method for isolation of human pancreatic islets. Diabetes 37:413–420

Scharp DW, Lacy PE, Santiago JV, McCullough CS, Weide LG, Falqui L, Marchetti P, Gingerich RI, Jaffe AS, Cryer PE, Anderson CB, Flye MW (1990) Insulin independence after islet transplantation into type I diabetes patients. Diabetes 39: 515–516

Sibley RK, Sutherland DER, Goethe F, Michael T (1985) Recurrent diabetes mellitus in the pancreas iso- and allografts. Lab Invest 53:132–140

Sutherland DER, Steffes MW, Mauer SM, Najarian JS (1974) Isolation of human and porcine islets of Langerhans and islet transplantation in pigs. J Surg Res 16: 102–111

Sutton R, Gray DWR, McShane P (1987) The metabolic efficiency and long-term fate of intraportal islet grafts in the cynomolgous monkey. Transplant Proc 19:3525–3576

Tzakis AG, Ricordi C, Alejandro R, Zeng Y, Fung JJ, Todo S, Demetris AJ, Mintz DH, Starzl TE (1990) Pancreatic islet transplantation after upper abdominal exenteration and liver replacement. Lancet 336:402–405

Warnock GL, Cattral MS, Rajotte RV (1988) Normoglycemia after implantation of purified islet cells in dogs. Can J Surg 31:421–426

Warnock GL, Ao Z, Cattral MS, Dabbs KD, Rajotte RV (1992a) Experimental islet transplantation in large animals. In: Ricordi C (ed) Pancreatic islet cell transplantation. Landes, Georgetown, pp 261–278

Warnock GL, Kneteman NM, Ryan EA, Rabinovitch A, Rajotte RV (1992b) Long-term follow-up after transplantation of insulin producing pancreatic islets into patients with type 1 (insulin-dependent) diabetes mellitus. Diabetologia 35:89–95

Younoszai R, Sorenson RL, Lindall AW (1970) Homotransplantation of isolated pancreatic islets. Diabetes 19: 406 (abstract)

7 Potential Use of Marrow Stromal Cells for Therapy of Osteogenesis Imperfecta and Osteoporosis

D.J. Prockop

The presence of hematopoietic stem cells (HSCs) in bone marrow has long been recognized (Hay 1966). Less well recognized is the fact that bone marrow also contains cells that meet the criteria for stem cells of nonhematopoietic tissues. The stemlike cells for nonhematopoietic tissues are currently referred to as either mesenchymal stem cells because of their ability to differentiate into cells that can roughly be defined as mesenchymal or as marrow stromal cells (MSCs) because they appear to arise from the complex array of supporting structures found in marrow. Isolated MSCs can readily be differentiated into osteoblasts, chondrocytes, adipocytes, and myoblasts in culture. Therefore, they are an intriguing model for defining the homology of stem cells. In addition, many of their properties appear to make them an ideal vehicle for cell and gene therapy.

The presence of stem cells for nonhematopoietic cells in bone marrow was first suggested by the German pathologist Cohnheim over 130 years ago (Cohnheim 1867). Cohnheim examined wound repair by injecting an insoluble analine dye into the veins of animals and then looking for the appearance of cells labeled with the dye in wounds he created at a distal site. Cohnheim concluded that most, if not all, of the cells that appeared in the wound came from the blood stream and by implication, from the bone marrow. The cells he identified included not only inflammatory cells but also cells with fibroblastlike morphology associated with thin fibrils. Therefore, he raised the possibility that bone marrow may be the source of fibroblasts for wound repair. Cohnheim's proposal of 1867 was tested in over 40 subsequent publications (see Ross et al. 1970; Davidson 1992). Most of the reports suggested the fibroblasts are of local origin, but the issue is still being examined (Bucala et al. 1994).

The first direct evidence for the presence of nonhematopoietic stem cells in marrow came from the work of Friedenstein beginning in the mid-1970s (Friedenstein et al. 1976). Friedenstein placed samples of whole bone marrow in plastic culture dishes and, after 4 h or so, poured off the majority of the cells that were nonadherent. He observed the small number of adherent cells were heterogeneous in appearance but the most characteristic cells appeared to be fibroblastlike cells that formed foci of two to four cells. The foci remained dormant for about 3 days and then began to replicate rapidly. The adherent cells became more uniformly fibroblastlike after passage several times. The most intriguing characteristic of the cells was that they differentiated into colonies that resembled small deposits of bone and cartilage under some culture conditions. Friedenstein's observations were subsequently repeated and extended by a number of investigators (see particularly Piersma et al. 1983; Beresford et al. 1992; Owen and Friedenstein 1988; Wakitani et al. 1995). The results thoroughly confirmed Friedenstein's observations and demonstrated that plastic-adherent cells from marrow were multipotential and readily differentiated into osteoblasts, chondroblasts, adipocytes, and even myoblasts. Friedenstein et al. (1987) further demonstrated that even after 20 or 30 cell doublings in culture, MSCs still had the potential to differentiate into bone and some cartilage when enclosed in a capsule with a porous membrane and implanted into the peritoneum of animals.

In spite of the extensive literature on MSCs, there is no consensus as to the phenotypic markers and protocols for isolation of the cells. Several sophisticated protocols have been reported (see Clark and Keating 1995; Long et al. 1995; Gronthos and Simmons 1996; Haynesworth et al. 1996; Waller et al. 1995; Rickard et al. 1996), but none have as yet been employed by more than one or two laboratories, and it has not been demonstrated that they all isolate the same cell populations.

My own research group tried to explore the question of where MSCs go after systemic infusion into animals (Pereira et al. 1995). Earlier reports (Piersma et al. 1983; van Zant et al. 1990; Anklesaria et al. 1989; Wu and Keating 1991) demonstrated that infused MSCs repopulate up to one third of the MSCs in marrow of recipient animals that have undergone marrow ablation by X-ray or chemotherapy to create a space for engraftment. We, however, were primarily interested in whether the infused MSCs repopulate nonhematopoietic cells and tissues. We isolated MSCs from mice as described by Friedenstein and others (Friedenstein et al. 1976, 1995; Piersma et al. 1983; Beresford et al. 1992; Owen and Friedenstein 1988; Wakitani et al. 1995) from a line of transgenic mice expressing a mutated collagen gene (Sokolov et al. 1993) that served as a convenient marker gene. The mutated collagen gene we had previously shown was expressed in a tissue-specific manner in transgenic mice. We, therefore, used it to serve both as a marker for the presence of donor cells and for the tissue-specific expression of any progeny of the donor cells that contained the marker gene.

One week after infusion into isogenic X-ray irradiated mice (Pereira et al. 1995), we did not detect any donor MSCs or progeny of MSCs in any of the tissues of recipient mouse with a polymerase chain reaction (PCR) assay that was sensitive enough to detect one donor cell per 10 000 cells (Fig. 1). After 1–5 months, however, the progeny of the donor MSCs accounted for about 5% of the cells in total tissues. Surprisingly, the extent of cell replacement was about the same in bone and cartilage as in marrow and spleen. The marker gene was found in bone cells that were cultured from cleanly dissected pieces of bone, and the mutated collagen was expressed as mRNA in the cultured bone cells. The gene was present but not expressed in chondrocytes from recipient mice. Since the marker gene of type I collagen is normally expressed in bone but not in cartilage (Sokolov et al. 1993), the results were consistent with tissue-specific expression of the marker gene by progeny of the MSCs. Subsequently, Keating et al. (1996) reported on similar experiments in which they detected donor MSCs or progeny of donor MSCs in liver, thymus and lung, in addition to marrow and spleen after infusion of human MSCs into nonirradiated SCID mice.

In our opinion, the simplest explanation for both the observations from our laboratory (Pereira et al. 1995) and by those of Keating et al. (1996) is that the donor MSCs first enter the bone marrow of recipient mice. Subsequently, they participated in the normal biological cycle in which MSCs in the bone marrow serve as a continuing source of cells for a variety of mesenchymal tissues. Since the marker gene for type I collagen in our experiments was expressed in bone but not in cartilage (Pereira et al. 1995), the progeny of MSCs apparently acquire the phenotypes of different target tissues either as they leave the marrow or after they have entered the microenvironment of the tissue itself.

It is now apparent that MSCs provide a highly intriguing source of a vehicle for both cell therapy and gene therapy. The MSCs are relatively easy to isolate from aspirates of bone marrow that can be obtained under local anesthesia. They are relatively easy to expand in culture and to transfect with exogenous genes. Hence, they have many apparent advantages over hematopoietic stem cells or other stem cells that have been tried as vehicles for cell and gene therapy (Schuening 1992). We have been involved in a collaborative project with Dr. Malcolm Brenner and Dr. Edwin Horwitz of St. Jude Childrens Research Hospital for the use

---

Fig. 1a,b. Distribution of progeny of donor marrow stromal cells (MSCs) after ▶ infusion into X-irradiated mice (Pereira et al. 1995). Recipients were 8- to 10-week-old mice from an inbred strain (FVB/N) that each received potentially lethal X-irradiation (9.0 Gy) before intravenous infusion of 1×10^5 cultured MSCs from a transgenic mouse with a mutated human COL1A1 gene together with 6×10^5 freshly isolated nonadherent cells from a normal mouse from the same strain as a source of hematopoietic cells. The mutated human COL1A1 gene and the endogenous mouse Col1a1 were assayed with a polymerase chain reaction (PCR) assay that used the same two primers for both genes but that generated a smaller fragment from the mouse Col1a1 gene because of a natural deletion in the 5-nontranslated region. After 1 week, the marker human COL1A1 gene was not detected in tissues from the recipient mice even though the assay would have detected 1:10 000 donor cells. At 1 month and 5 months, progeny of the donor MSCs accounted for 1%–12% of the cells in marrow, spleen, bone, lung, and cartilage. Because of the relatively large variation in the data, the values at 1 month and 5 months are not significantly different. **a** Assay of ^{32}P-labeled PCR products obtained with DNA isolation from bone as a template and separated on a 7% polyacrylamide-6 M urea gel. The gel was assayed with a phosphostimulatable storage plate (PhosphorImager; Molecular Dynamics, Sunnyvale, CA). **b** Values for several tissues expressed as percent of donor MSCs or progeny of donor MSCs per total cells. Symbols: – and +, mice that did not or did receive X-irradiation before infusion of donor MSCs plus nonadherent cells; *1–3* and *A–C*, arbitrary notations assigned to mice killed at 1 month or 5 months. *H*, PCR product from the human marker gene consisting of a mutated COL1A1 gene; *M*, PCR product from the endogenous mouse COL1A1

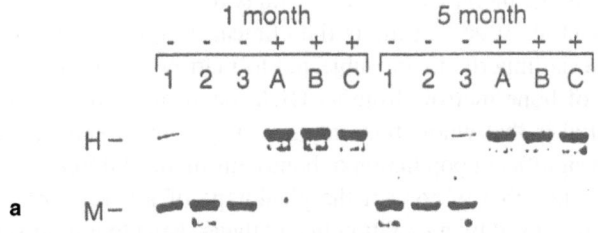

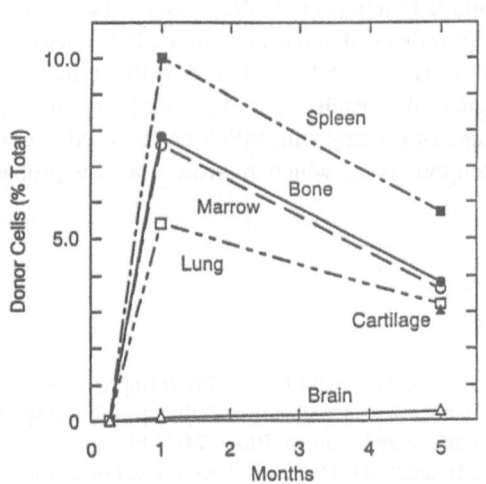

Fig. 1a,b. Legend see p. 114

of MSCs in the therapy of children with severe osteogenesis imperfecta (Horwitz et al. 1996). In the initial clinical trial, children with severe osteogenesis imperfecta are subjected to marrow ablation and then an infusion of bone marrow from an HLA-identical donor. The rationale for the trial is that whole bone marrow may contain enough MSCs to replace a significant population of bone cells in the children. As a result, it may be possible to convert the phenotype of a severe osteogenesis imperfecta caused by a mutation in a collagen gene to a mild variant of the same disease. In the future, we and others plan to isolate MSCs from patients with osteogenesis imperfecta, correct the mutated genes in culture, and return the corrected MSCs to the same patient. Several reports by others (Bienzle et al. 1994; Quesenbery et al. 1994) suggest that if large numbers of bone marrow cells are used for transfusion, engraftment of HSC can be achieved without the need for marrow ablation. Hence, the results raise the possibility that, in the future, a similar strategy of therapy with MSCs can be used in common diseases such as osteoporosis in which marrow ablation probably cannot be justified.

References

Anklesaria P, Kase K, Glowacki J et al. (1989) Improved hematopoiesis in anemic S1/S1d mice by spenectomy and therapeutic transplantation of a hematopoietic microenvironment. Blood 74:1144

Beresford JN, Bennett JH, Devlin C, Leboy PS, Owen ME (1992) Evidence for an inverse relationship between the differentiation of adipocytic and osteogenic cells in rat marrow stromal cell cultures. J Cell Sci 102:341

Bienzle D, Abrams-Ogg AC, Kruth SA, Ackland-Snow J, Carter RF, Dicke JE, Jacobs RM, Kamel-Reid S, Dube ID (1994) Gene transfer into hematopoietic stem cells: long-term maintenance of in vitro activated progenitors without marrow ablation. Proc Natl Acad Sci USA 91:350

Bucala R, Spiegel LA, Chesney J, Hogan N, Cerami A (1994) Circulating fibrocytes define a new leukocyte subpopulation that mediates tissue repair. Molec Med 1:71

Clark BR, Keating A (1995) Biology of bone marrow stroma. Ann NY Acad Sci 770:70

Cohnheim J (1867) Über Entzundung und Eiterung. Arch Path Anat Physiol Klin Med 40:1

Davidson JM (1992) Wound repair. In: Gallin JI, Goldstein IM, Snyderman R
    (eds) Inflammation: basic principles and clinical correlates, 2nd edn. Raven,
    New York, pp 809–819
Friedenstein AJ, Chailakhyan RK, Gerasimov UV (1987) Bone marrow osteo-
    genic stem cells: in vitro cultivation and transplantation in diffusion cham-
    bers. Cell Tissue Kinet 20:263
Friedenstein AJ, Gorskaja U, Kulagina NN (1976) Fibroblast precursors in
    normal and irradiated mouse hematopoietic organs. Exp Hematol 4:276
Gronthos S, Simmons PJ (1996) The biology and application of human bone
    marrow stromal cell precursors. J Hematother 5:15
Hay E (1966) Regeneration. Holt, Rinehart and Winston, New York
Haynesworth SE, Baber MA, Caplan AI (1996) Cytokine expression by human
    marrow-derived mesenchymal progenitor cells in vitro: effects of dex-
    amethasone and IL-1a. J Cell Physiol 166:585
Horwitz EM, Prockop DJ, Marini J, Fitzpatrick LA, Pyeritz R, Sussman M,
    Orchid P, Brenner MK (1996) Treatment of severe osteogenesis imperfecta
    by allogeneic bone marrow transplantation. Matrix Biol 15:188
Keating A, Guinn B, Larava P, Wang XH (1996) Human marrow stromal cells
    electrotransferred with human factor IX (FIX) cDNA engraft in SCID
    mouse marrow and transcribe human FIX. Exp Hematol 24:180
Long MW, Robinson JA, Ashcraft EA, Mann KG (1995) Regulation of human
    bone marrow-derived osteoprotenitor cells by osteogenic growth factors. J
    Clin Invest 95:881
Owen ME, Friedenstein AJ (1988) Cell and molecular biology of vertebrate
    hard tissues. Ciba Found Symp 136:42–60
Pereira RF, Halford KW, OHara MD, Leeper DB, Sokolov BP, Pollard MD,
    Bagasra O, Prockop DJ (1995) Cultured adherent cells from marrow can
    serve as long-lasting precursor cells for bone, cartilage, and lung in irradi-
    ated mice Proc Natl Acad Sci USA 92:4857
Piersma AH, Ploemacher RE, Brockbank KGM (1983) Transplantation of
    bone marrow fibroblastoid stromal cells in mice via the intravenous route.
    Br J Haematol 54:285
Piersma AH, Ploemacher RE, Brockbank KG (1983) Transplantation of bone
    marrow fibroblastoid stromal cells in mice via the intravenous route. Br J
    Haematol 54:285
Quesenbery PJ, Crittenden RB, Lowry P, Kittler EW, Ras S, Peters S, Ram-
    shaw H, Stewart FM (1994) Engraftment of normal murine marrow into
    nonmeyloablated host mice. Blood Cells 20:348
Rickard DJ, Kassem M, Hefferan TE, Sarkar G, Spelsberg TC, Riggs BL
    (1996) Isolation and characterization of osteoblast precursor cells from hu-
    man bone marrow. J Bone Min Res 11:312

Ross R, Everett NB, Tyler R (1970) Wound healing and collagen formation VI. The origin of the wound fibroblast studied in paraboisis. J Cell Biol 44:645

Schuening FG (1992) Gene transfer into hematopoietic stem cells. Curr Top Microbiol Immunol 177:237

Sokolov BP, Mays PK, Khillan JS, Prockop DJ (1993) Tissue- and development-specific expression in transgenic mice of a type I procollagen (COL1A1) minigene construct with 2.3 kb of the promoter region and 2 kb of the 3-flanking region. Specificity is independent of the putative regulatory sequences in the first intron. Biochemistry 32:9242

van Zant G, Holland BP, Eldridge PW, et al. (1990) Genotype-restrictied growth and aging patterns in hematopoietic stem cell populations of allophenic mice. J Exp Med 1:1547

Wakitani S, Saito T, Caplan AI (1995) Myogenic cells derived from rat bone marrow mesenchymal stem cells exposed to 5-azacytidine. Muscle Nerve 18:1412

Waller EK, Huang S, Terstappen L (1995) Changes in the growth properties of CD34+, CD38- bone marrow progenitors during human fetal development. Blood 86:710

Wu DD, Keating A (1991) Engraftment of donor derived bone marrow stromal cells. Exp Hematol 19:485 (abstr)

8 Neuronal Modulation of the Immune Response in Nervous Tissue: Implications for Neurodegenerative and Autoimmune Diseases

H. Neumann and H. Wekerle

8.1 Introduction

This review summarizes recent data highlighting an interconnection between the nervous and the immune system which may be relevant for developing an effective therapy for neurodegenerative or autoimmune diseases. Traditionally, neurodegenerative diseases have been considered as local nervous system processes with no conceivable pathogenic link to inflammatory or autoimmune diseases of the nervous system. However, this view does not explain the local immune response that occurs in lesions of primary neurodegenerative processes. This response is characterized by up-regulation of major histocompatibility complex (MHC) molecules and immune-like infiltrations. One example is amyotrophic lateral sclerosis, which in some cases appears to be triggered by a point mutation in the superoxide dismutase gene (Andersen

et al. 1995). Lymphocytes and monocytes form prominent infiltrates and
MHC molecules are induced on microglia in areas of the central nervous
system (CNS) affected by degeneration (Lampson et al. 1990;
Kawamata et al. 1992). Similar signs of local immune activation are
detectable in other neurodegenerative processes such as Parkinson's or
Alzheimer's disease. In the latter, CD4 and CD8 T lymphocytes infiltrate the lesioned areas and microglia cells are activated to attain antigen-presenting properties. In the degenerative lesions microglial cells
are stimulated to express MHC molecules, which likely present antigens
to T lymphocytes (Itagaki et al. 1988; Kalaria 1993).

Recent experimental data suggest that there are complex interactions
between degenerative and inflammatory brain diseases and may explain
the close association of neurodegenerative processes and local immune
responses. It appears that both neurons and T cells participate in controlling the immune reactivity of the nervous tissue. Proinflammatory cytokines derived from T cells activate MHC genes for antigen presentation, while electrically active neurons counteract this induction. Some of
the evidence which has led to this concept will be reviewed below.

8.2 The Immune Response in the Nervous System

Immune responses are initiated and executed through cognate interactions of T cell receptors on T lymphocytes with MHC molecules on
target cells. Typically, intracellular peptides such as viral antigens or
auto-antigens are presented by MHC class I molecules to CD8+ T cells.
Phagocytosed and processed proteins are presented by MHC class II on
specialized antigen-presenting cells (APCs). In conjunction with MHC
molecules, adequately stimulated APCs express costimulatory factors, a
process which facilitates antigen-specific activation of T lymphocytes.
These cells subsequently initiate a local inflammatory response and
attain effector function, e.g., cytotoxicity, upon contact with a prestimu-
lated APC expressing the particular T cell-specific peptide. These pres-
timulated APCs do not only activate T cells recognizing peptides derived from microbial or viral proteins, they also can transform
autoreactive T cell clones, which are by no means always depleted from
the normal immune repertoire, to auto-aggressive T cells with cytotoxic
potential (Wekerle et al. 1996). MHC products for antigen presentation

are expressed in most tissues with the exception of the healthy CNS, which appears to be an immune privileged organ with no resident antigen presenting function. Furthermore, the nervous tissue is separated from the circulating immune cells by a blood-brain barrier. This physical barrier, which is formed by a network of endothelial cell tight junctions, controls the penetration of macromolecules such as immunoglobulins and the immigration of immune cells in the CNS parenchyma. Nevertheless the CNS is an immune-competent organ and the integrity of the brain is strictly controlled by the immune system (Wekerle et al. 1986). Much of our knowledge of the brain's immune status which will be described below was derived from studies using experimental allergic encephalomyelitis as a model system (Wekerle et al. 1994). Activated, but not resting, T cells pass the blood-brain barrier irrespective of their antigen receptor specificity (Hickey et al. 1991). They traffic through the nervous tissue parenchyma and are presumably responsible for immune surveillance of brain tissue. When activated, these T cells secrete proinflammatory cytokines which induce glia cells, especially microglia and astrocytes, to produce MHC antigens and other molecules to productively present protein (auto-)-antigen to the immune cells (Wekerle 1994). Induction of MHC molecules, the antigen presenting processing machinery and costimulatory molecules is the prerequisite for presentation of local (auto-)-antigen and for activation of T cells. No immune response can be initiated or accelerated by trafficking T lymphocytes in the healthy brain, because their target structures, i.e., MHC molecules on APCs, are lacking and are not inducible. As will be discussed below, the proinflammatory stimuli produced by T cells either must be: (1) extremely strong and long-lasting, overwhelming the suppressive mechanisms which prevent MHC expression in the healthy CNS or (2) associated with a neuronal lesion which disturbs local control of the inflammatory reaction. These brain-infiltrating T cells appear, at least in part, to be eliminated if not required or after completion of their task by programmed cell death.

8.3 Neuronal Lesion and Immune Activation

Several years ago an unexpected link between neuronal injury and immune reactivity in the CNS became obvious. Several independent groups reported that transection of peripheral nerves leads to a retrograde response of microglia cells adjacent to the lesioned neuronal somata. Microglia cells become activated and deafferentiate the lesioned motorneurons by displacement of the input synaptic terminals (Blinzinger and Kreutzberg 1968). These perineuronal glia cells respond with de novo expression of immunologically relevant membrane proteins including MHC class I and class II molecules and LFA-1 adhesion molecules (Maehlen et al. 1988; Streit et al. 1989; Moneta et al. 1993). Induction of these proteins rendered the activated glia cells competent for interaction with immunosurveillant T lymphocytes. As mentioned above, activated T cells migrate into the brain and carry out immune surveillance of the tissue parenchyma. Do they also find and seek out a neurodegenerative process ? In fact, reactivated myelin protein-specific T cells injected intravenously are attracted to the severed optic nerve, resulting in infiltration along the affected optic tract but not on the contralateral intact side (Hickey 1991; Konno et al. 1990). Similar results came from experiments with facial nerve transection in rats with experimental allergic encephalomyelitis. These animals developed elevated T cell infiltration and marked signs of inflammation within the facial nucleus of the peripherally lesioned nerve, but not in the contralateral intact nucleus (Maehlen et al. 1989).

The exact cellular mechanisms of this neuronal-glial interaction leading to up-regulation of MHC and adhesion molecules on adjacent glia cells and consequently to T cell chemoattraction are still obscure. One would expect that glia cells activated to express MHC molecules have acquired all functions for taking up, processing and presenting protein antigens in the context of MHC class I or class II molecules. Thus migrating T lymphocytes can potentially recognize peptides from foreign proteins, in the case of infection, or their autoantigenic epitope from genuine brain structures.

The experimental data of these studies suggest that impaired neuronal function or neuronal degeneration in response to an aseptic neuronal injury leads to activation of glia cells, mainly microglia cells. Production of MHC molecules is stimulated on these microglia cells, which then become APCs and thus acquire the capacity of specifically interacting with immigrating T lymphocytes.

8.4 Neuronal Regulation of MHC Expression

Lack of MHC expression in the healthy CNS is not due to the inability of brain cells to express the relevant genes. In fact, it has previously been established that most brain cell types are capable of expressing MHC molecules upon appropriate stimulus. In vitro pro-inflammatory cytokines like interferon-γ (IFN-γ) induce expression of MHC class I or MHC class II molecules in most glia lineage (Wong et al. 1984). Lack of MHC expression in the CNS thus should be due either to a deficit of inducing factors or to active suppressive control. Evidence for neuronal control of MHC expression comes from recent studies examining MHC class II expression of astrocytes and microglia in hippocampal explant culture (Neumann et al. 1996). These cultured slices, derived from neonatal rat hippocampal tissue, differentiate in about 5–7 days and mimic the organotypic neuronal-glial organization found in vivo (Gähwiler 1988). The cultured neurons are interconnected to synaptic circuits and show spontaneous action potentials.

Immune fluorescence analyses of these organotypic hippocampal cultures revealed that MHC class II inducibility of astrocytes and microglia was low or undetectable in the central areas of a slice culture harboring an intact neuronal architecture. Astrocytes and microglia located in areas that are devoid of neurons were readily susceptible to IFN-γ-mediated MHC class II induction. The striking difference in cytokine responsiveness is presumably controlled by neurons. This is supported by results showing that suppression of MHC class II inducibility is completely lost after inhibition of neuronal activity with the sodium channel blocker tetrodotoxin. Following application of tetrodotoxin, astrocytes are induced to express MHC class II throughout the entire slice culture and microglial cells exhibit a higher level of MHC class II expression. Tetrodotoxin treatment of pure astrocyte or microglia cultures did not interfere with MHC class II inducibility. These data strongly suggest that electrically active neurons suppress cytokine-mediated MHC class II induction on adjacent glia cells.

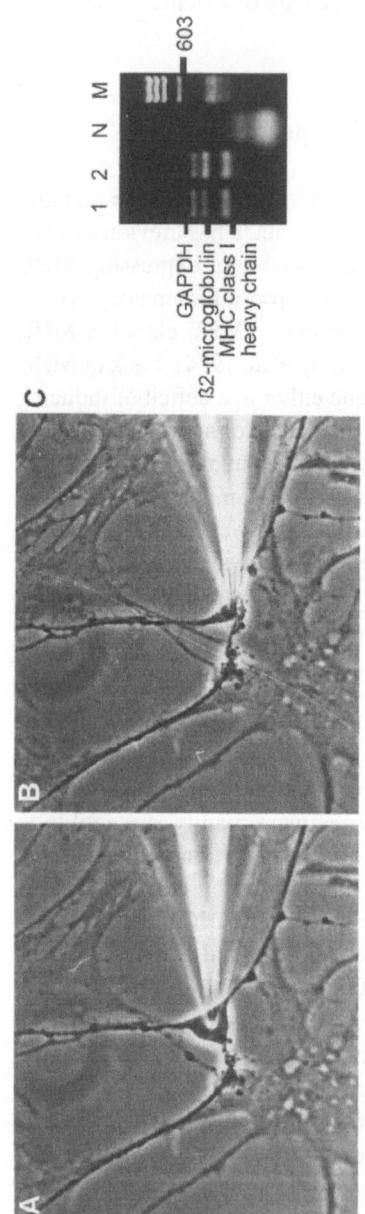

Fig. 1A–C. Legend see p. 125

MHC molecules are not only down-regulated by neurons in an activity-dependent fashion on neighboring astrocytes and microglia, but also within the neuronal cells themselves. Until very recently, it was debated whether neurons have a primary deficiency of MHC expression or whether MHC genes could be induced under certain stimuli. Recent evidence demonstrates that MHC class I genes are very tight regulated in neurons, but these cells are capable of producing and expressing MHC class I molecules (Neumann et al. 1995). Neuronal transcription of MHC class I genes was analyzed by a combination of patch-clamp electrophysiology with single-cell reverse transcriptase polymerase chain reaction (RT-PCR) (Lambolez et al. 1992; Monyer and Lambolez 1995). A sample of the cytoplasm was harvested with a micropipette directly after electrophysiological characterization of the actual functional state of the neuron (Fig. 1A, B). Cytoplasmic RNA was collected, reverse transcribed and analyzed by PCR to assess the genes currently transcribed (Fig. 1C). The gene transcripts for β2-microglobulin and the peptide transporter TAP1/TAP2 (both proteins required for correct MHC class I assembly) were lacking in neurons with very few exceptions. Surprisingly, however, the MHC class I heavy chain was constitutively transcribed in a considerable proportion of cultured rat hippocampal neurons. IFN-γ was capable of inducing MHC class I and MHC class I-related genes in a small proportion of neurons, which, in contrast to the majority of the cells, failed to show spontaneous electric activity. Consistent with previous experiments of glial MHC inducibility, paralysis of neurons with tetrodotoxin resulted in a significant increase of β2-microglobulin and TAP gene transcription, coincident with MHC class I protein expression on the neuronal membrane.

◀ **Fig. 1A–C.** Single-cell RT-PCR of MHC class I genes in hippocampal neurons. **A** Dissociated rat hippocampal neurons were characterized by whole-cell patch-clamp recording. **B** Directly after electrophysiology the cytolasm was harvested through the micropipette. Gene transcripts for the housekeeping enzyme GAPDH were coamplified to control the quality of the cytoplasmic RNA sample. **C** Cytoplasmic RNA was reverse transcribed and multiplex PCR was performed. MHC class I light chain (β2-microglobulin) and heavy chain were detectable by RT-PCR in individual neurons (*lanes 1, 2*) after treatment with interferon-γ and tetrodotoxin. *M*, molecular weight marker; *N*, negative PCR control

The data show that the MHC class I-related genes (TAP and β2-microglobulin) are very tightly regulated in neurons, but that MHC class I molecules can definitely be expressed on the cell surface under certain circumstances. Effective induction of MHC class I is achieved by the proinflammatory cytokine IFN-γ in neurons lacking sodium-dependent activity.

8.5 Conclusions

Antigen presentation, which facilitates immune response, is inhibited in healthy brain tissue, a finding which leads us to conclude that neurons control this exceptional immune status in the nervous system. We suggest that tight regulation of MHC expression in the brain has developed to protect the sensitive and irreplaceable neurons against damage produced by inflammatory reactions. Negative regulation of MHC inducibility in neurons and adjacent glia cells would exempt functional neurons in intact brain tissue from an (auto-)immune attack because of their lack of MHC expression. In contrast, lesioned tissue, with damaged neurons, would become more easily inducible and upon MHC expression could be recognized and eventually attacked by cytotoxic T cells.

Neuronal control of nervous tissue immune reactivity appears to be impaired in diseases involving neuronal loss or decreased synaptic innervation, both typical features of neurodegenerative disorders. Consequently, neurodegenerative diseases of the CNS are associated with local activation of the immune response. The function of the impressive changes in local immune reactivity, with up-regulation of MHC molecules in areas of neuronal degeneration, is difficult to interpret. It is not known at present whether the local inflammatory reaction accelerates the neurodegenerative process or whether it reflects a yet to be understood helpful function of the immune system within the brain.

References

Andersen PM, Nilsson P, Ala-Hurula V, Keränen M-J, Tarvainen I, Haltia T, Nilsson L, Binzer M, Forsgren L, Marklund SL (1995) Amyotrophic lateral sclerosis associated with homozygosity for an $Asp^{90}Ala$ mutation in CuZn-superoxyde dismutase. Nature Genet 10:61–65

Blinzinger K, Kreutzberg GW (1968) Displacement of synaptic terminals from regenerating motoneurons by microglial cells. Z Zellforsch 86:145–157

Gähwiler BH (1988) Organotypic cultures of neural tissue. Trends Neurosci 11:484–489

Hickey WF (1991) Migration of hematogenous cells through the blood-brain barrier and the initiation of CNS inflammation. Brain Pathol 1:97–106

Hickey WF, Hsu BL, Kimura H (1991) T lymphocyte entry into the central nervous system. J Neurosci Res 28:254–260

Itagaki S, McGeer PL, Akiyama H (1988) Presence of T-cytotoxic suppressor and leukocyte common antigen-positive cells in Alzheimer's disease brain tissue. Neurosci Lett 91:259–264

Kalaria RN (1993) The immunopathology of Alzheimer's disease and some related disorders. Brain Pathol 3:333–347

Kawamata T, Akiyama H, Yamada T, McGeer PL (1992) Immunologic reactions in amyotrophic lateral sclerosis brain and spinal cord. Am J Pathol 140:691–707

Konno H, Yamamoto T, Suzuki H, Yamamoto H., Iwasaki Y, Ohara Y, Terunuma H, Harata N (1990) Targeting of adoptively transferred experimental allergic encephalomyelitis lesion at the site of Wallerian degeneration. Acta Neuropathol 80:521–526

Lambolez B, Audinat E, Bochet P, Crépel F, Rossier J (1992) AMPA receptor subunits expressed by single Purkinje cells. Neuron 9:247–258

Lampson LA, Kushner PD, Sobel RA (1990) Major histocompatibility complex antigen expression in the affected tissues in amyotrophic lateral sclerosis. Ann Neurol 28:365–372

Maehlen J, Schröder HD., Klareskog L, Olsson T, Kristensson K (1988) Axotomy induces MHC class I expression on rat nerve cells. Neurosci Lett 92:8–13

Maehlen J, Nennesmo I, Olsson A-B, Olsson T, Schröder HD, Kristensson K (1989) Peripheral nerve injury causes transient expression of MHC class I antigens in rat motor neurons and skeletal muscle. Brain Res 481:368–372

Moneta ME, Gehrmann J, Töpper R, Banati RB, Kreutzberg GW (1993) Cell adhesion molecule expression in the regenerating rat facial nucleus. J Neuroimmunol 45:203–206

Monyer H, Lambolez B. (1995) Molecular biology and physiology at the single-cell level. Curr Opin Neurobiol 5:382–387

Neumann H. Boucraut J, Hahnel C., Misgeld T, Wekerle H (1996) Neuronal control of MHC class II inducibility in rat astrocytes and microglia. Eur J Neurosci 8: 2582–2590

Neumann H, Cavalié A, Jenne DE, Wekerle H (1995) Induction of MHC class I genes in neurons. Science 269:549–552

Streit WJ, Graeber MB, Kreutzberg GW (1989) Peripheral nerve lesion produces increased levels of major histocompatibility complex antigens in the central nervous system. J Neuroimmunol 21:117–123

Wekerle H, Linington C, Lassmann H, Meyermann R (1986) Cellular immune reactivity within the CNS. Trends Neurosci 9:271–277

Wekerle H (1994) Antigen presentation by CNS glia. In: Kettenmann H, Ransom B (eds) Neuroglial cells. Oxford University, Oxford, pp685–699

Wekerle H, Kojima K, Lannes-Vieira J, Lassmann H, Linington C (1994) Animal models. Ann Neurol 36:S47-S53

Wekerle H, Bradl M, Linington C, Kääb G, Kojima K (1996) The shaping of the brain-specific T lymphocyte repertoire in the thymus. Immunol Rev 149:231–243

Wong GHW, Bartlett PF. Clark-Lewis I, Battye F, Schrader JW (1984) Inducible expression of H-2 and Ia antigens on brain cells. Nature 310:688–691

9 Transplantation of Dopamine-Synthesizing Cells in Parkinson's Disease: Experimental Data and Clinical Application

W.H. Oertel, O. Pogarell, A. Kupsch, and A. Hartmann

9.1 Introduction

Transplantation of fetal dopaminergic cells into the striatum as an experimental therapeutic strategy for Parkinson's disease (PD) was introduced in 1979 by Björklund and Stenevi, and Perlow et al. In 1987, the first PD patient was transplanted according to this procedure (Lindvall et al. 1989). Since then, substantial preclinical and clinical data suggest that neurotransplantation can be a potentially effective treatment for PD, especially for patients who, 5–10 years after onset of symptoms, develop serious therapy-related side effects such as marked motor fluctuations including end-of-dose akinesia, ON-OFF-phenomena, dystonia, and/or dyskinesia (Freed et al. 1992; Freeman et al. 1995a; Peschanski et al. 1994). Major topic-related issues that have been solved as well as unresolved questions will be outlined in the following. A special effort will be made to demonstrate mutual interrelations between experimental and clinical research in this field.

9.2 Mechanism of Clinical Benefit in PD

Transplanted dopaminergic neurons provide an endogeneous source of dopamine, thereby reducing or even abolishing the need for exogeneously administered dopamimetic compounds. In favor of a dopamine related benefit are the following observations: Clinical improvement after striatal grafting is mainly contralateral. It increases gradually over weeks to months, has been sustained for more than 3 years and can be correlated with an increase in striatal fluorodopa (FD)-uptake in positron emission tomography (PET) scans (Lindvall et al. 1990, 1994; Kordower et al. 1995). However, additional factors may contribute to the clinical benefit observed in experimental models of PD following transplantation or in PD patients who received a fetal graft.

First, transplanted fetal mesencephalic cells not only secrete dopamine, but are further able to store exogenously administered l-DOPA-derived dopamine and to buffer dopamine by re-uptake, both of which potentially reduce central side effects related to concomitant pharmacologic dopaminergic therapy. Second, additional release of trophic factors by transplanted cells of dopaminergic or nondopaminergic origin needs to be taken into consideration, especially with regard to remaining nigral neurons prone to degeneration. Third, lesion-induced release of neurotrophic factors could be a contributing factor, although the amount of fetal tissue transplanted quite clearly correlates with clinical benefit; further, use of large-bore needles is associated with a diminished clinical response, although they cause increased tissue trauma (Lindvall et al. 1992). In addition, sham-transplanted 4-methyl-1-phenyl-1,2,3,6-tetrahydropyridine (MPTP) monkeys fail to display behavioral improvement following surgery (Taylor et al. 1995). Fourth, a placebo effect could play a role in the clinical response observed, since transplanted patients are subject to increased care and evaluation. To rule out, or rather quantify, the amount of clinical benefit likely to be related to a potential placebo effect, a double-blind, placebo-controlled study (transplantation vs. sham operation consisting of a burr-hole) is currently under way in the United States. Should fetal transplantation be associated with a significant clinical improvement, the placebo-treated patients will also receive fetal transplants at the end of the 16-month follow-up period.

9.3 Type of Tissue

Early studies utilizing autotransplants of adrenal medullary have not yielded sufficient long-term clinical benefit in animals or in PD patients (Freed et al. 1990; Peterson et al. 1989). This finding was in accordance with no or hardly any survival of chromaffin cells, as confirmed by autopsies (Hirsch et al. 1990; Kordower et al. 1991). Moreover, this procedure involved a considerable mortality rate of about 10%. Therefore, this technique was abandoned by most centers, although in theory, substantial advantages could initially be anticipated with respect to lack of immunological rejection of the graft and the availability of donor tissue as the latter is donored by the patient/host (Goetz et al. 1991). However, recent evidence from laboratory studies suggest that co-application of trophic factors by adding peripheral nerve tissue to adrenal medullary grafts (as to fetal mesencephalic grafts) may significantly improve survival of the transplantated dopaminergic neurons (Collier et al. 1994; Unsicker 1993). Whether these findings will eventually bring about a renaissance for autologous adrenal medullary grafts remains to be determined (Bakey 1995; Date al. 1995).

Presently, the use of fetal mesencephalic allografts represents the tissue of choice in dopaminergic cerebral transplantation and will therefore be discussed in more detail.

9.4 Issues in Intrastriatal Tranplantation of Fetal Mesencephalic Tissue for PD

In 1997, more than 300 patients worldwide have undergone intrastriatal transplantation of fetal mesencephalic tissue with published follow-up periods of 46 months maximum (Olanow et al. 1996). As a result, several issues have been clarified.

9.4.1 Donor Age and Source

It is currently agreed that the optimal donor age is between 5.5 and 9 weeks postconception (PC). The rationale is that during this period, dopaminergic cells first appear in the mesencepahlic subventricular layer, further differentiate, and extend neuritic processes. This time window was assessed by in vivo studies investigating the viability of human ventral mesencephalic humans in immunosuppressed rodents (Freeman et al. 1995b). This finding is also supported by clinical results which show that good survival of implanted nigral neurons is only achieved when tissue is exclusively derived from donors in the optimal age window (Freeman et al. 1995a; Kordower et al. 1995). In vitro, it appears that suspension grafts optimally survive when cells are obtained between 5.5 and 8 weeks PC, whereas for solid grafts, this interval is between 6.5 and 9 weeks (Freeman et al. 1995b).

Another issue relates to the source of donor tissue. In a rodent model of PD, fetal tissue obtained from spontaneous abortions has resulted in functional benefit (Kondoh et al. 1994). However, tissue obtained from elective abortions is more likely devoid of possible genetic defects, uninfected and viable, since fetal death precedes spontaneous abortions in 50% of instances (Freeman and Olanow 1991).

9.4.2 Solid Versus Suspension Grafts

In animal models as well as in PD patients, both graft types have shown clinical benefit (Lindvall et al. 1994; Freed et al. 1992; Freeman et al. 1995a). Solid grafts, as mentioned earlier, extend the window for viable grafting by 1 week (Freeman et al. 1991). In addition, they are technically easier to prepare and display normal cytoarchitecture. In contrast, suspension grafts are more homogeneous and, theoretically, offer the possibility of purified dopaminergic cell implants. However, by pooling multiple donors, rejection or infection of one cell line can result in the loss of all deposits. So far, no consensus has emerged as to which graft type is preferable (Olanow et al. 1996).

9.4.3 Number of Donors

At least 200 000 nigral neurons projecting to the putamen degenerate in PD (Pakkenberg et al. 1991). Therefore, complete replacement therapy would require between five and ten donors, since 20 000–40 000 dopaminergic neurons per donor survive transplantation in animal models (Kordower et al. 1996). A recent postmortem analysis of a patient who had been transplanted with six embryonic grafts 18 months prior to his death and clinically showed substantial improvement revealed more than 200 000 viable neurons as assessed by tyrosine hydroxylase (TH) immunoreactivity (Kordower et al. 1995). However, it is unclear how many transplanted cells actually establish functional contacts within the striatum. In general, clinical and PET scan studies indicate that there is a positive correlation between the number of donors and subsequent clinical response (Remy et al. 1995). Ideally, a dose–response study should be conducted to establish the optimal number of donors for fetal striatal grafting in PD patients.

9.4.4 Age and Disease State

Current indications for surgical PD therapy (including a functional neurosurgical procedure such as lesion or high-frequency stimulation of the globus pallidus internus, subthalamic nucleus, or ventralis intermedius nucleus) comprise patients with advanced PD who experience severe medication-associated side effects, i.e., motor fluctuations (Oertel and Quinn 1996). However, it is theoretically conceivable that early PD patients might also represent a suitable group for fetal striatal transplantation. One reason is the better tolerability of the surgical procedure compared to elder, potentially multimorbid patients. More particularly, it can be speculated that early transplanted PD patients might bypass long-term medication-related adverse effects, which make grafting necessary in the first place. In addition, avoidance of l-DOPA therapy may improve graft survival, since l-DOPA has been claimed to have toxic effects on grafted neurons (Steece-Collier et al. 1995). Ultimately, this issue will have to be clarified in long-term prospective studies.

9.4.5 Site of Implantation

Initially, unilateral grafts were used for intrastriatal transplantation of fetal dopaminergic cells. This procedure resulted in mainly unilateral symptom improvement in parkinsonian animals as in PD patients (Annett et al. 1995; Lindvall et al. 1990; Peschanski et al. 1994). This observation was supported by PET scan studies demonstrating increased FD uptake on the operated side, whereas, contralaterally, a decline in FD uptake was noted, indicating an ongoing degeneration of striatonigral dopaminergic neurons (Lindvall et al. 1994; Sawle et al. 1992). Therefore, it is plausible to assume that bilateral grafting will result in better clinical results and is now the procedure of choice in most ongoing cerebral transplantation trials in PD patients.

Concerning the precise site of implantation, there is increasing evidence from animal studies to suggest that, besides transplantation into the putamen, where most of the degenerating nigral neurons project to, the caudate nucleus may also be an important target region (Annett et al. 1995). In humans, combined grafting into the putamen/caudate is increasingly being used (New York, USA/Lund, Sweden), and preliminary data suggest a good subsequent clinical response.

As to the distribution of grafted tissue, 5-mm spaces appear to be the minimum to completely reinnervate the striatum, as suggested by laboratory data and confirmed in humans using PET and autopsy studies to determine the amount of reinnervation post-grafting (Freeman et al. 1995a; Kordower et al. 1995). Microtransplantation, which consists in an even denser surface coverage of the striatum, has shown to be beneficial in a rat model of PD (Nikkah et al. 1994). Whether microtransplantation will result in improved clinical outcome in humans remains to be determined.

9.4.6 Immunosuppression

Conflicting data exist on whether immunosuppression is necessary in cerebral transplantation. Fetal allografts in rodents and nonhuman primates have been shown to survive for substantial periods of time without immunosuppression (Björklund 1992). Moreover, donor-specific sensitization or nonspecific immunological responses have not been observed following intrastriatal grafting in nonhuman primates; there is one published case of a PD patient who has been successfully transplanted without any concomitant immunosuppression (Freed et al. 1992; Henderson et al. 1991). In contrast, allograft rejection has been observed in rodents as well as in one autopsied PD patient (Folkerth and Durso 1996; Hudson et al. 1994). Therefore, transplantation of fetal allografts into humans usually involves a regimen of at least 6 months of cyclosporin A postoperatively (Kordower et al. 1995). At present, it is unclear whether immunosuppression can be withheld entirely. A current placebo-controlled, prospective study in PD patients in the USA is, if possible, actually going to refrain from concomitant administration of immunosuppressant agents. In contrast, in an ongoing trinational BIOMED II program (Lund/Sweden; London, UK; Marburg, Germany) a triple regimen with corticosteroids, azathioprin, and cyclosporin A is administered for at least 12 months.

9.4.7 Tissue Storage

Currently, fetal mesencephalic tissue can be stored for up to 1 week in human amnionic fluid, thereby allowing transportation of tissue, screening for infectious agents, acquisition of multiple donors, and elective scheduling of surgery (Freed et al. 1992). The use of cryopreservation has not proven to be a useful method yet, as determined in in vivo and in vitro studies (Frodl et al. 1994; Sauer et al. 1992; Sautter et al. 1996). In particular, this method is associated with significantly reduced viability and neuritic outgrowth of transplanted neurons (Collier et al. 1993). However, it would certainly be desirable to develop a method which allows storage of fetal mesencepahlic tissue for months or even years while still achieving satisfactory graft viability. In future work, methods need to be developed with which this goal can be achieved if cerebral transplantation is ever going to be used on a routine basis clinically.

9.4.8 Supporting Strategies

There are conclusive data from animal and human studies that formation and reduced elimination of free radicals play a crucial role in the pathogenesis of PD (Coyle and Puttfarcken 1993; Fahn and Cohen 1992). Recent animal data have shown that genetically manipulated mice cells hyperexpressing the scavenger enzyme Cu/Zn dismutase (Nakao et al. 1995) or the concomitant use of 21-aminosteroids (Grasbon-Frodl et al. 1996; Nakao et al. 1994) with antioxidant properties (lazaroids) significantly improve viability of grafted dopaminergic neurons in vitro. Moreover, the same effect could be reproduced in vivo, as assesssed by graft survival and functional recovery, when these cells were transplanted into 6-hydroxydopamine (OHDA)-lesioned rats (Nakao et al. 1994, 1995). Since systemic administration of lazaroids is safer and technically easier than transplantation of cells derived from transgenic animals, human studies are currently underway in which this compound is being admnistered concomitantly (BIOMED II program).

Also, the use of neurotrophic factors as a potential strategy to improve graft survival (and possibly for the rescue of dopaminergic neurons prone to degeneration in the disease course) appears to offer promising perspectives in PD patients. In particular, glial cell line-derived neurotrophic factor (GDNF) has been shown to be the most specific neurotrophin for dopaminergic neurons in vitro and in vivo (Beck et al. 1995; Gash et al. 1996; Tomac et al. 1995). To date, with respect to animal models of PD, it is the only substance known to be able to rescue dopaminergic neurons after intrastriatal 6-OHDA lesions in the rat (Sauer et al. 1995). However, blood–brain barrier passage of systemically administered GDNF is probably not sufficient to induce therapeutic effects. The possibility of producing genetically engineered cells secreting GDNF has been recently described (Sagot et al 1996). Whether transplantation of such cells will yield positive results in animal models of PD remains to be determined. Apart from intracerebroventricular (i.c.v.) administration, intraparenchymal implantation of GDNF-producing cells appears to be the most promising venue for use of neurotrophic factors in humans.

9.4.9 Clinical Assessment

In order to facilitate comparison of clinical results among transplant centers in Europe and the USA, the "Core Assessment Protocol for Intracerebral Transplantation" (CAPIT; CAPIT Committee 1991) has been developed. Furthermore, many clinical protocols include neuropsychological testing, quality of life and self-imaging assessment (Sass et al. 1995). In addition, objective measures of clinical improvement can be determined by electrophysiological parameters. Finally, striatal FD uptake measured by PET has become the standard procedure in the evaluation of striatal dopaminergic function. Another important issue concerns the ethical problems linked to cerebral transplantation. The "Network of European CNS Transplantation and Restoration" (NECTAR) has set up a criteria list for minimal ethical standards to be met within the associated European cerebral transplant centers (Boer 1994).

9.5 Open Questions and Future Perspectives

The availability of fetal mesencephalic tissue is, should intracerebral transplantation eventually become a more widely used procedure for PD patients, a very realistic problem from a logistical as well as from an ethical perspective. Therefore, alternatives are already under investigation. All are based on xenotransplantational strategies.

First, implantation of genetically engineered polymer-encapsulated PC12 cells, where dopamine and other substances, i. e., growth factors, can diffuse along a semipermeable membrane, has been investigated in animal models of PD (Aebischer et al. 1994, 1991a; Tresco et al. 1992;

Winn et al. 1991). One major advantage appears to be that these cell compounds can be explanted or exchanged whenever necessary. Moreover, they offer the possibility of grafting cross-species cells without inducing immune rejection (Aebischer et al. 1991b).

So far, functional recovery following capsule implantation has only been described in animals models of parkinsonism. A preliminary study in humans, in which hamster kidney cells producing ciliary neurotrophic factor (CTNF) have been implanted in amyotrophic lateral sclerosis (ALS) patients, has failed to demonstrate any clinical benefit, although capsules explanted after a period of 4–6 months contained adequate amounts of viable cells (Aebischer et al. 1996). The question remains whether substance delivery i.c.v. or even in target areas is generally sufficient to induce satisfactory clinical effects. More specificly, one of the major requirements for successful neural transplantation may depend on the graft's ability to form functional contacts with preexisting CNS neurons. For example, GDNF applied next to the substantia nigra in 6-OHDA-lesioned rats was able to rescue degenerating nigral neurons; however, these surviving neurons were unable to form functional contacts with remaining striatal neurons, i. e., to induce axonal sprouting or form synaptic contacts, resulting in an only modest symptomatic improvement (Winkler et al. 1996).

Second, the use of xenografts could represent a potential source of fetal tissue. The major issue in this respect is obviously an immunological one. Possibly, masking of major histocompatibility complex (MHC-I) expression in xenografts may diminish the likelihood of rejection (Pakzaban et al. 1995). Four patients transplanted with fetal mesencephalic neurons derived from pigs with concomitant immunosuppression have tolerated the procedure so far without serious side effects (O. Isacson, personal communication). However, the clinical response seems to be diminished compared to the use of human fetal tissue.

Third, progenitor cells derived from the subependymal layer of rats and mice may represent a further source of transplantation tissue. These cells can be easily cultured, can be genetically engineered to release different substances, and are able to form functional contacts with neurons adjacent to the target area, as in vivo studies have indicated (Franklin et al. 1995; Lacorazza et al 1996; Martinez-Serrano et al. 1995). However, their use in humans remains to be established, especially with regard to their xenogeneic nature.

9.6 Summary

In summary, important information derived from animal data have already enabled a therapeutic use of intracerebral transplantation in – though few – humans with PD. At present, this therapy for PD is still experimental. Whether its use in patients will be offered on a larger scale deserves further investigation. This depends on at least three issues: the availability of donor tissue, the agreement on ethical guidelines by society, and finally, the results in cerebral transplantation will have to be compared with those of new pharmacotherapeutic approaches and functional neurosurgery.

References

Aebischer P, Tresco PA, Sagen J, Winn SR (1991a) Transplantation of microencapsulated bovine chromaffin cells reduces lesion-induced rotational asymmetry in rats. Brain Res 560:43–49

Aebischer P, Tresco PA, Winn SR, Greene LA, Jaeger CB (1991b) Long-term cross-species transplantation of a polymer-encapsulated dopamine-secreting cell line. Exp Neurol 111:269–275

Aebischer P, Goddard M, Signore AP, Timpson RL (1994) Functional recovery in hemiparkinsonian primates transplanted with polymer-encapsulated PC12 cells. Exp Neurol 126:151–158

Aebischer P, Schluep M, Deglon N, Joseph JM, Hirt L, Heyd B, Goddard M, Hammang M, Zurn AD, Kato AC, Regli F, Baetge EE (1996) Intrathecal delivery of CTNF using encapsulated genetically modified xenogeneic cells in amyotrophic lateral sclerosis patients. Nat Med 2:696–699

Annett LE, Torres EM, Ridley RM, Baker HF, Dunnett SB (1995) A comparison of the behavioural effects of embryonic nigral grafts in the caudate nucleus and in the putamen of marmosets with unilateral 6-OHDA lesions. Exp Brain Res 103:355–371

Bakey RAE (1995) Stereotactic intrastriatal co-grafts of autologuous adrenal medulla (AM) and peripheral nerve (PN) improves motor performance in Parkinson's disease [comment]. Neurosurgery 37:518–519

Beck KD, Valverde J, Alexi T, Poulsen K, Moffat B, Vandlen RA, Rosenthal A, Hefti F (1995) Mesencephalic dopaminergic neurons protected by GDNF from axotomy-induced degeneration in the adult brain. Nature 26:339–341

Björklund A, Steveni U (1979) Reconstruction of the nigrostriatal dopamine pathway by intracerebral nigral transplants. Brain Res 177:555–560

Björklund A (1992) Dopaminergic transplants in experimental Parkinsonism: cellular mechanisms of graft-induced functional recovery. Curr Opin Neurobiol 2: 683–689

Boer G (on behalf of NECTAR) (1994) Ethical guidelines for the use of human embryonic or fetal tissue for experimental and clinical neurotransplantation and research. J Neurol 242:1–13

CAPIT Committee (1991) Core assessment program for intracerebral transplantations. In: Lindvall O, Björklund A, Widner H (eds) Intracerebral transplantation in movement disorders. restorative neurology. (vol 4) Elsevier, Amsterdam-Tokyo, pp232–241

Collier TJ, Gallagher MJ, Sladek CD (1993) Cryopreservation and storage of embryonic rat mesencephalic dopamine neurons for one year: comparison to fresh tissue in culture and neural grafts. Brain Res 623:249–256

Collier TJ, Elsworth JD, Taylor JR, Sladek JR Jr, Roth RH, Redmond DE Jr (1994) Peripheral nerve-dopamine neuron co-grafts in MPTP-treated monkeys: augmentation of tyrosine hydroxylase-positive fiber staining and dopamine content in host systems. Neuroscience 61:875–889

Coyle JT, Puttfarcken P (1993) Oxidative stress, glutamate, and neurodegenerative disorders. Science 262:689–695

Date I, Asari S, Ohmoto T (1995) Two-year follow-up study of a patient with Parkinson's disease and severe motor fluctuations treated by co-grafts of adrenal medulla and peripheral nerve into bilateral caudate nuclei: case report. Neurosurgery 37:515–519

Fahn S, Cohen G (1992) The oxidant stress hypothesis in Parkinson's disease: evidence supporting it. Ann Neurol 32:804–812

Folkerth RD, Durso R (1996) Survival and proliferation of nonneural tissues, with obstruction of cerebral ventricles, in a parkinsonian patient treated with featl allografts. Neurology 46:1219–1225

Franklin RJ, Bayley SA, Milner R, Ffrench-Constant C, Blakemore WF (1995) Differentiation of the O-2A progenitor cell line CG-4 into oligodendrocytes and astrocytes following transplantation into glia-deficient areas of CNS white matter. Glia 13:39–44

Freed WJ, Poltorak M, Becker JB (1990) Intracerebral medullary grafts: a review. Exp Neurol 110:139–166

Freed CR, Breeze RE, Rosenberg NL, Schneck SA, Kriek E, Qi JX, Lone T, Zhang YB, Snyder JA, Wells TH et al (1992) Survival of implanted fetal dopamine cells and neurologic improvement 12 to 46 months after transplantation for Parkinson's disease. Ann Neurol 327:1549–1555

Freeman TB, Olanow CW (1991) Fetal homotransplants in the treatment of Parkinson's disease. Arch Neurol 48: 900–902

Freeman TB, Olanow CW, Hauser RA, Nauert GM, Smith DA, Borlogan CV, Sanberg PR, Holt DA, Kordower JH, Vingerhoets FJG, Snow BJ, Calne D,

Gauger LL (1995a) Bilateral fetal nigral transplantation into the postcommissural putamen in Parkinson's disease. Ann Neurol 38:379–378

Freeman TB,Sanberg PR, Nauert GM, Boss BD, Spector D, Olanow CW, Kordower JH (1995b) The influence of donor age on the survival of solid and suspension intraparenchymal human embryonic nigral grafts. Cell Transplant 4:141–154

Freeman TB, Spence MS, Boss BD, Spector DH, Strecker RE, Olanow CW, Kordower JH (1991) Development of dopaminergic neurons in the human substantia nigra. Exp Neurol 113: 344–353

Frodl EM, Duan WM, Sauer H, Kupsch A, Brundin P (1994) Human embryogenic dopamine neurons xenografted to the rat: effects of cryopreservation and varying regional source of donor cells on transplant survival, morphology and function. Brain Res 647:286–298

Gash DM, Zhang Z, Ovadia A et al (1996) Functional recovery in parkinsonian monkeys treated with GDNF. Nature 380:252–255

Goetz CG, Stebbins GT, Klawans HL, Koller WC, Grossman RF, Bakey RAE, Penn RD (1991) United Parkinson Foundation neurotransplantation registry on adrenal medullary transplants: presurgical and 1-year and 2-year follow-up. Neurology 41:1719–1722

Grasbon-Frodl EM, Nakao N, Brundin P (1996) The lazaroid U-83836E improves the survival of rat embryonic mesencephalic tissue stored at 4°C and subsequently used for cultures or intracerebral transplantation. Brain Res Bull 6:341–346

Henderson BT, Clough CG, Hughes RC, Hitchcock ER, Kenny BG (1991) Implantation of human fetal ventral mesencephalon to the right caudate nucleus in advanced Parkinson's disease. Arch Neurol 48:822–827

Hirsch EC, Duyckaerts C, Javoy-Agid F, Hauw JJ, Agid Y (1990) Does adrenal graft enhance recovery of dopaminergic neurons in Parkinson's disease? Ann Neurol 27: 676–682

Hudson JL, Hoffman A, Stromberg I, Hoffer BJ, Moorhead JW (1994) Allogeneic grafts of fetal dopamine neurons: behavioral indices of immunological interactions. Neurosci Lett 171:32–36

Kondoh T, Blount JP, Conrad JA, Pundt LL, Low WC (1994) Functional effects of transplanted human fetal ventral mesencephalic brain tissue from spontaneous abortions into a rodent model of Parkinsons disease. Transplant Proc 26: 3335

Kordower JH, Cochran E, Penn RD, Goetz CG (1991) Putative chromaffin cell survival and enhanced host-derived TH-fiber innervation following a functional adrenal medulla autograft for Parkinson's disease. Ann Neurol 29: 405–412

Kordower JH, Freeman TB, Snow BJ, Vingerhoets FJG, Mufson EJ, Sanberg PR, Hauser RA, Smith DA, Nauert GM, Perl DP, Olanow CW (1995)

Neuropathological evidence of graft survival and striatal reinnervation after the transplantation of fetal mesencephalic tissue in a patient with Parkinson's disease. N Engl J Med 332:1118–1124

Kordower JH, Rosenstein JM, Collier TC, Burke MA, Chen EY, Li JM, Martel L, Levey AE, Mufson EJ, Freeman TB, and Olanow CW (1996) Functional fetal nigral grafts in Parkinson's disease: chemoanatomic, ultrastructural and metabolic studies. J Comp Neurol 370:203–230

Lacorazza HD, Flax JD, Snyder EY, Jenoubi M (1996) Expression of human beta-hexosaminidase alpha-subunit gene (the gene defect for Tay-Sachs disease) in mouse brains upon engraftment of transduced progenitor cells. Nat Med 2:424–429

Lindvall O, Rehncrona S, Brundin P, Gustavii B, Astedt B, Widner H, Lindholm T, Björklund A, Leenders KL, Rothwell JC, Frackowiak R, Marsden CD, Johnels B, Steg G, Freedman R, Hoffer BJ, Seiger A, Bygdeman M, Stromberg I, Olson L (1989) Human fetal dopamine neurons grafted into the striatum in two patients with severe Parkinson's disease. A detailed account of methodology and a 6-month follow-up. Arch Neurol 46:615–631

Lindvall O, Brundin P, Widner H, Rehncrona S, Gustavii B, Frackowiak R, Leenders KL, Sawle G, Rothwell JC, Marsden CD et al (1990) Grafts of fetal dopamine neurons survive and improve motor function in Parkinsons's disease. Science 247:574–577

Lindvall O, Widner H, Rehncrona S, Brundin P, Odin P, Gustavii B, Frackowiak R, Leenders KL, Sawle G, Rothwell JC, Björklund A, Marsden CD (1992) Transplantation of fetal dopamine neurons in Parkinson's disease: one-year clinical and neurophysiological observations in two patients with putaminal implants. Ann Neurol 31:155–165

Lindvall O, Sawle G, Widner H, Rothwell JC, Björklund A, Brooks D, Brundin P, Frackowiak R,Marsden CD, Gustavii B, Odin P, Rehncrona S (1994) Evidence for long-term survival and function of dopaminergic grafts in progressive Parkinson's disease. Ann Neurol 35:172–180

Martinez-Serrano A, Fischer W, Björklund A (1995) Reversal of age-dependent cognitive impairments and cholinergic neuron atrophy by NGF-secreting neural progenitors grafted to the basal forebrain. Neuron 15:473–484

Nakao N, Frodl EM, Duan WM, Widner H, Brundin P (1994) Lazaroids improve the survival of grafted rat embryonic dopaminergic neurons. Proc Natl Acad Sci USA 91:12408–12412

Nakao N, Frodl EM, Widner H, et al (1995) Overexpressing Cu/Zn-superoxide dismutase enhances survival of transplanted neurons in a rat model of Parkinson's disease. Nat Med 1:226–231

Nikkah G, Cunningham MG, Jödicke H, Knappe U, Björklund A (1994) Improved graft survival and striatal teinnervation by microtransplantation of fetal nigral cell suspensions in the rat Parkinson model. Brain Res 687:133–143

Oertel WH, Quinn NP (1996) Parkinsonism. In: Brandt T, Caplan LR, Dichgans J, Diener CH, Kennard C (eds) Neurological disorders: course and treatment. Academic, San Diego, pp 715–772

Olanow CW, Kordower JH, Freeman TB (1996) Fetal nigral transplantation as a therapy for Parkinson's disease. Trends Neurosci 19:102–109

Pakkenberg B, Moller A, Gundersen HJ, Mouritzen-Dam A, Pakkenberg H (1991) The absolute number of nerve cells in substantia nigra in normal subjects and in patients with Parkinson's disease estimated with an unbiased stereological method. J Neurol Neurosurg Psychiatry 54:30–33

Pakzaban P, Deacon TW, Burns LH, Dinsmore J, Isacson O (1995) A novel mode of immunoprotection of neural xenotransplants: masking of donor histocompatibility complex class I enhances transplant survival in the central nervous system. Neuroscience 65:993–996

Perlow MI, Freed WJ, Hoffer BJ, Seiger A, Olson L, Wyatt RJ (1979) Brain grafts reduce motor abnormalities produced by destruction of the nigrostriatal dopamine system. Science 204:643–647

Peschanski M, Defer G, N'Guyen JP, Ricolfi F, Montfort JC, Remy P, Geny C, Samson Y, Hantraye P, Jeny R et al (1994) Bilateral motor improvement and alteration of L-dopa effect in two patients with Parkinson's disease following intrastriatal transplantation of foetal ventral mesecephalon. Brain 117:487–499

Peterson DI, Price ML, Small CS (1989) Autopsy findings in a patient who had an adrenal-to-brain transplant for Parkinson's disease. Neurology 39: 235–238

Remy P, Samson Y, Hantraye P, Fontaine A, Defer G, Mangin JF, Fénelon G, Gény C, Ricolfi F, Frouin V, N'Guyen JP, Jeny R, Degos JD, Peschanski M, Cesaro P (1995) Clinical correlates of (18F) fluorodopa uptake in five grafted parkinsonian patients. Ann Neurol 38:580–588

Sagot Y, Tan SA, Hammang JP, Aebischer P, Kato AC (1996) GDNF loss of motoneurons but not axonal degeneration or premature death of pmn/pmn mice. J Neurosci 16:2335–2341

Sass KJ, Buchanan CP, Westerveld CP, Marek KL, Fan A, Robbins A, Naftolin F, Vollmer TL, Leranth C, Roth RH, Price LH, Bunney BS, Elsworth JD, Hoffer PB, Redmond DE, Spencer DD (1995) General cognitive ability following unilateral and bilateral fetal ventral mesencephalic tissue transplantation for treatment of Parkinson's disease. Arch Neurol 52:680–686

Sauer H, Frodl EM, Kupsch A, ten Bruggencate G, Oertel WH (1992) Cryopreservation, survival and function of intrastriatal fetal mesencephalic grafts in a rat model of Parkinson's disease. Exp Brain Res 90:54–62

Sauer H, Rosenblad C, Björklund A (1995) GDNF but not TGF-3β prevents delayed degeneration of nigral dopaminergic neurons following striatal 6-hydroxydopamine lesion. Proc Natl Acad Sci USA 92:8935–8939

Sautter J, Strecker S, Kupsch A, Oertel WH (1996) Methylcellulose during cryopreservation of ventral mesencephalic tissue fragments fails to improve survival and function of cell suspension grafts. J Neurosci Meth 64: 173–179

Sawle GV, Bloomfield PM, Björklund A, Brooks DJ, Brundin P, Leenders KL, Lindvall O, Marsden CD, Rehncrona S, Widner H et al (1992) Transplantation of fetal dopamine neurons in Parkinson's disease: PET [18F]6-L-fluorodopa studies in two patients with putaminal implants. Ann Neurol 31:166–173

Steece-Collier K, Yurek DM, Collier TJ, Junn FS, Sladek JR Jr (1995) The detrimental effect of levodopa on behavioral efficacy of fetal dopamine neuron grafts in rats is reversible following prolonged withdrawal of chronic dosing. Brain Res 676:404–408

Taylor JR, Elsworth JD, Sladek JR Jr, Colliet TJ, Roth RH, Redmond DE Jr (1995) Sham surgery does not ameliorate MPTP-induced behavioral deficits in monkeys. Cell Transplant 4:13–26

Tomac A, Lindquist E, Lin LF, Ogren SO, Young D, Hoffer BJ, Olson L (1995) Protection and repair of the nigrostriatal system by GDNF in vivo. Nature 373:335–339

Tresco PA, Winn SR, Aebischer P (1992) Polymer encapsulated neurotransmitter secreting cells. Potential treatment for Parkinson's disease. ASAIO J 38:17–23

Unsicker K (1993) The trophic cocktail made by adrenal chromaffin cells. Exp Neurol 123:167–173

Winkler C, Sauer H, Lee CS, Björklund A (1996) Short-term GDNF treatment provides long-term rescue of lesioned nigral dopaminergic neurons in a rat model of Parkinson's disease. J Neurosci 16:7206–7215

Winn SR, Tresco PA, Zielinski B, Greene LA, Jaeger CB, Aebischer P (1991) Behavioral recovery following intrastriatal implantation of microencapsulated PC12 cells. Exp Neurol 113:322–329

10 Repairing CNS Demyelination by Glial Cell Transplantation

R.J.M. Franklin

10.1 Introduction

Primary demyelination following oligodendrocyte injury or death is unusual in the spectrum of pathological events that occur in the CNS in that it may be followed by spontaneous regeneration. This regeneration, in which saltatory conduction is restored to demyelinated axons by re-investing them with new myelin sheaths (Smith et al. 1979), is termed remyelination and can lead to functional recovery (Jeffery and Blakemore 1996). Remyelination has been well documented in a variety of experimental models of demyelination (Ludwin 1988) and undoubtedly occurs in naturally occurring demyelinating diseases, such in the acute phase of multiple sclerosis (MS) (Prineas et al. 1993a). Were complete remyelination to occur in all circumstances, then demyelinating diseases would carry a favourable prognosis with a strong likelihood of recovery. Unfortunately, remyelination is not a reliable process and there are many occasions where it fails or is not sustained. For example, the remyelination that follows experimentally induced demyelination in old animals is very poor compared with the robust repair that occurs in young animals (Gilson and Blakemore 1993), while in MS, foci of demyelination that fail to undergo remyelination characterise the later stages of the disease (Prineas et al. 1993b). The reasons why remyelination fails are unclear and are likely to remain so until we have a better understanding of the cellular and molecular basis by which remyelination proceeds. Indeed, a lack of understanding of the basic mechanisms of remyelination has hindered the development of rational approaches to re-activating or enhancing spontaneous remyelination. However, the development of glial cell transplantation as an experimental technique to study cellular interactions during glial development and repair has spawned the idea that transplantation of myelinogenic cells might be used to remyelinate areas of persistent demyelination in clinical conditions (Blakemore and Franklin 1991; Franklin 1993; Duncan 1996; Franklin and ffrench-Constant 1996). This article will review recent advances in experimental models of glial cell transplantation and critically relate these to the possible development of transplantation as a treatment of human demyelinating disease. The term glial cell transplantation will mostly be used to refer to transplantation of myelinogenic glial cells rather than astrocytes, about which there is a growing literature (Blakemore et al. 1995b).

10.2 Experimental Models for Glial Cell Transplantation

An important feature of all glial cell transplantation models is that the host environment into which cells are transplanted contains nonmyelinated axons that would be myelinated under normal circumstances. Such an environment can arise for a variety of reasons, and many have been used as host environments in transplantation studies, such as the nonmyelinated axons of the retina (Blakemore et al. 1986; Huang et al. 1991) or during development before myelination is complete (Espinosa de los Monteros et al. 1993; Freidrich and Lazzarini 1993; Lachapelle et al. 1994). However, the majority of transplantation studies have been undertaken using one of two models.

The first of these are the so-called myelin-mutants in which the CNS never becomes fully myelinated due to defects in the genes encoding one of the major myelin proteins. Of these, the hypomyelinated shiverer mouse (an MBP mutant) has been widely used since the pioneering studies of Gumpel and her co-workers at the Hôpital de la Salpêtrière in Paris, France. The mutation not only results in there being axons available for myelination by the transplanted cells but also provides the basis for identification of transplant-derived myelin sheaths which are MBP+ve and, in contrast to the mutant, contain a major-dense line (Lachapelle et al. 1984; Gumpel et al. 1989; Warrington et al. 1993). Shaking pup (Archer et al. 1997) and particularly the myelin-deficient rat (Duncan et al. 1988; Rosenbluth et al. 1990; Archer et al. 1994; Tontsch et al. 1994) (both X-linked PLP mutants) have also been used for transplantation studies on account of their profound hypomyelination.

The second model involves the creation of focal areas of demyelination in adult white matter by direct injection of ethidium bromide (Blakemore and Crang 1988; Franklin et al. 1991) or lysolecithin (Vignais et al. 1993; Baron-Van Evercooren et al. 1996), both of which are agents toxic to oligodendrocytes. Since gliotoxin-induced demyelination undergoes spontaneous remyelination to a greater or lesser degree, these models provide a situation where transplanted cells interact with host-derived cells during repair. A particularly useful variant of the gliotoxin model developed by Blakemore and colleagues involves creating demyelination in tissue exposed to 40 Gy of X-irradiation. This dose of X-irradiation has the effect of suppressing the spontaneous repair response, resulting in a demyelination lesion which is never repopulated by host glia and which fails to remyelinate. Such a lesion provides an excellent environment with which to ask questions about the behaviour and, particularly, the myelinogenic properties of transplanted glia (Blakemore and Crang 1985, 1988; Groves et al. 1993; Trotter et al. 1993; Barnett et al. 1993a; Franklin et al. 1995, 1996).

Although none of these models accurately mimic the chronic demyelinated plaques of MS, they have nevertheless provided valuable information from which one can assess the likely suitability of a range of transplanted cell populations.

10.3 What Are the Properties Required of a Transplanted Cell for Repairing Areas of Demyelination?

Clearly, the ability to generate new myelin sheaths capable of restoring saltatory conduction to demyelinated axons is a principal requirement of the transplanted cells. However, two further properties are likely to be beneficial. Firstly, the ability of the transplanted cell to undergo controlled proliferation would enable the transplanted population to generate sufficient cells appropriate to the extent of the area needing repair in response to local mitogens. Although proliferation alone may constitute a means by which transplanted cells can fill a lesion, the ability of the transplanted cells to migrate is likely to assist this process. Moreover, in a multifocal condition such as MS it would also be beneficial if cells were not only able to migrate within but also between areas of demyelination. If this were the case, then multiple injections of cells, in which each clinically relevant plaques is targeted, might not be necessary. With these three requirements in mind, one can examine the suitability of the available types of myelinogenic cell.

10.4 The Oligodendrocyte Lineage

The oligodendrocyte lineage has been well characterised in tissue culture studies and as a result it has been possible to isolate populations of cells enriched for a particular phenotypic stage which can then be transplanted into one of the available transplantation models. It is clear from such studies that all in vitro stages of the oligodendrocyte lineage (OL), from the pre-A2B5+ progenitor to O1+ oligodendrocyte, are able to myelinate axons following transplantation (Frankin et al. 1993; Groves et al. 1993; Duncan et al. 1992; Warrington et al. 1993). The ability of mature myelinating oligodendrocytes isolated from adult tissue to myelinate axons for a second time following transplantation is more controversial (Ludwin and Szuchet 1993; Targett et al. 1996). However, the various stages of the lineage differ in their proliferate and migratory capacity when transplanted into myelin deficient mutant models, where it is apparent that earlier stages generate more myelin over a wider area than do later stages (Warrington et al. 1993; Rosenbluth et al. 1990). This lineage variability following transplantation reflects well-documented differences between the biology of the motile, proliferative oligodendrocyte progenitor and the nonmotile, nondividing differentiated oligodendrocyte (Pfeiffer et al. 1993). Thus, according to the criteria for suitability outlined above, the oligodendrocyte progenitor seems the most suitable cell to use. A further advantage of the oligodendrocyte progenitor is that it may also be able to exhibit a differentiation plasticity that is not available to a mature oligodendrocytes. Under certain circumstances this cell may give rise to an astrocyte as well as a myelinating oligodendrocyte, reflecting a bipotentiality that can be demonstrated in vitro but appears not to occur to any great extent during normal development (Barnett et al. 1993a; Trotter et al. 1993; Franklin et al. 1995; Franklin and Blakemore 1995).

The ability of the transplanted oligodendrocyte progenitor to migrate through normal, intact adult CNS has been a matter of some controversy. It is clear that these cells can migrate following transplantation into developing normal or myelin-deficient CNS, where they presumably join the endogenous population of migrating cells (Gansmüller et al. 1991; Jacque et al. 1992; Friedrich and Lazzarini 1993; Lachapelle et al. 1994; Tontsch et al. 1994). A number of studies undertaken in the mouse have reported that oligodendrocyte progenitors will not only migrate through normal adult tissue but that they will also preferentially migrate towards areas of demyelination (Vignais et al. 1993; Gout and Dubois-Dalcq 1993). However, the findings of these studies are at variance with more recently published work which comes to a somewhat less optimistic conclusion (Franklin et al. 1996). When CG4 cells, a rat cell line widely used to mimic the behaviour of primary-derived oligodendrocyte, are transplanted into normal adult rat white matter none of the cells migrate and progressively fewer cells are found at the point of injection, suggesting poor survival. Only when the cells are placed close to active areas of demyelination are cells able to enter the lesion, and even then this is not a consistent event. If the cells are transplanted as little as a single intervertebral space from the area of demyelination, then none of the transplanted cells are able to traverse the intervening normal tissue. The reason for the poor survival of CG4 cells transplanted into normal adult tissue is not known, but it seems likely that they succumb to a shortage of survival factors, the levels of which are tightly regulated during development and adulthood to match the number of OL cells, initially generated in excess, to the number of

axons requiring myelination (Barres and Raff 1994). Presumably, the
levels of survival factor are elevated within active areas of demyelination (Komoly et al. 1992; Tourbah et al. 1992; McKay et al. 1996), since
when oligodendrocyte progenitors are transplanted directly into such an
environment they survive and are able to remyelinate demyelinated
axons. Since cells appear to survive when present at the edges of a
demyelinating lesion, one can imagine a scenario in which survival
factors present within the lesion are also present as a diminishing gradient around the lesion. The presence of penumbral effects around a
lesion, for which some evidence exists (Oumesmar et al. 1995), may
account for the apparent discrepancy between the observations in the rat
and those in the mouse, where the relative extent of the penumbra and
the precise position of the transplant with respect to the lesions borders
may lead to apparently opposite effects on the behaviour of transplanted
oligodendrocyte progenitor cells.

If one accepts that oligodendrocyte progenitors transplanted into
normal tissue fail to survive then the prospects for remyelinating widespread multifocal areas of demyelination by transplantation of a suppos-
edly migratory population of cells are greatly diminished. To overcome
this problem it will be necessary to change the properties (or identity) of
the transplanted cell and/or modify the environment in such a way that
it supports survival and migration. With regard to the latter it has been
shown that the nonpermissive adult CNS can be made supportive of
CG4 and primary oligodendrocyte progenitor survival, proliferation and
migration by prior exposure to 40 Gy of X-irradiation (Franklin et al.
1996; O'Leary and Blakemore 1997). Indeed, if the tissue is treated in
this way, then transplanted cells not only survive and migrate but will
also enter remote areas of demyelination – which they proceed to
remyelinate (Franklin et al. 1996). The nature of the changes in permis-
siveness brought about by X-irradiation are the subject of current research efforts. Clearly, if one could identify these factors, then it would
be possible to devise alternative means of altering the ability of adult
CNS to support transplanted oligodendrocyte progenitors that do not
carry the potentially hazardous side effects of high dose X-irradiation.

From where might one obtain oligodendrocyte progenitors for human transplantation? It is apparent from a number of studies that nonhis-
tocompatible glial transplants undergo rejection (Crang and Blakemore
1991; Archer et al. 1994; Rosenbluth et al. 1993), although this problem

could be side-stepped if one were to use cells obtained from the patient. The virtual absence of the subependymal plate in adult human, in contrast to the large and active plate in the adult rat, means that it is unlikely to be as rich a source of uncommitted progenitor cells. Nevertheless, a number of recent studies have shown that oligodendrocyte progenitors can be isolated from adult human brain, although the numbers of cells that can reasonably be obtained would not be sufficiently large to allow transplantation (Armstrong et al. 1992; Gogate et al. 1994; Scolding et al. 1995). Would it be possible, however, to expand these cell in vitro with appropriate combinations of growth factors as is possible using oligodendrocyte progenitors from the neonatal rat? Although these cells would appear capable of division, the identity of the mitogens remains elusive and it seems likely that they are different from those in the rat (Scolding et al. 1995). Until it is possible to expand oligodendrocyte progenitors from adult human brain, one must be cautious about the prospects of autologous transplantation.

10.5 Schwann Cells

Given that cells of the OL fail to fulfil all the requirements of a transplanted cell, one might look to the other myelinating cell of the nervous system, the Schwann cell. At first glance the case for the Schwann cell looks promising. These cells (1) myelinate demyelinated CNS axons, either by following transplantation (Blakemore and Crang 1985) or when recruited from endogenous sources (Blakemore 1978), and thereby restore saltatory conduction (Felts and Smith 1992; Honmou et al. 1996), (2) can be grown and expanded from autologous biopsy material (Rutkowski et al. 1995), and (3), unlike oligodendrocytes, would not be targets of autoimmune injury that underlies the genesis of the demyelinating lesions of MS. However, in spite of this impressive list of credentials, the interactions that occur between astrocytes and Schwann cells impose limitations on their suitability. There is a large body of evidence from developmental, pathological and in vitro studies which indicates that Schwann cells will only occupy regions of the nervous system from which astrocytes are absent (Franklin and Blakemore 1993; Baron-van Evercooren 1993; Ghirnikar and Eng 1994). They do not appear to migrate through astrocyte-defined territories, nor will they myelinate axons surrounded by astrocytic processes such as would occur in chronic MS plaques (Blakemore et al. 1986). It will be necessary to devise ways of overcoming the mutual exclusivity between Schwann cells and astrocytes based on a better understanding of its molecular basis before Schwann cell transplantation for MS becomes a feasible proposition.

10.6 Olfactory Bulb-Ensheathing Cells

The olfactory bulb-ensheathing cells are a distinct group of glia which ensheath the small diameter axons of olfactory nerve as it enters the olfactory bulb. Although found within the CNS, they share properties with both Schwann cells and astrocytes (Doucette 1990; Barnett et al. 1993b). We have recently shown that the Schwann cell-like properties of these cells can be pushed to the extent that these cells will attain a myelinating phenotype when confronted with demyelinated large diameter axons following transplantation into an X-irradiated ethidium bromide lesion in the adult rat spinal cord (Franklin et al. 1996b). This myelinating phenotype resembles a myelinating Schwann cell both ultrastructurally and in synthesising myelin-containing P0, a protein unique to peripheral myelin. It seems reasonable to suppose that these myelin sheaths will restore conduction much as Schwann cells are able to. However, the current interest in these cells is that, unlike Schwann cells, these cells co-exist with astrocytes in their normal environment and may therefore not be subject to the restrictions that a CNS glial environment imposes on the behaviour of Schwann cells. Moreover, it is conceivable that these cells could be obtained by autologous biopsy. The remyelination potential of these cells is currently under investigation.

10.7 Concluding Remarks

Although much progress has been made in the glial cell transplantation field in recent years, there are a number of key issues for which further information is required before one could be confident about transferring experimental approaches to clinical situations. Some of these issues have already been addressed, such as how one might improve the survival and migration of transplanted cells, the need to learn more about the biology of the human OL, and from where one might obtain a reliable source of transplant cells. There are a number of other issues which require careful consideration. For example, would transplant-repaired plaques be subjected to the same demyelinating process which led to the development of the initial lesion? If one where to use OL cells, then the answer is almost certainly yes, and the success of the approach would be dependent on the continuing advances that are being made in the prevention of initial demyelination. Then there is the issue of patient selection. Within the spectrum of clinical manifestations of MS, which category of patients is most likely to benefit? Since lesions may occur throughout the neuraxis, in which anatomical locations is transplantation most likely to achieve successful recovery?

Finally, it is worth bearing in mind that transplantation is only one of a number of possible approaches to promoting repair of CNS demyelination. Identification of growth factor mediators of inherent remyelination, coupled with improved techniques for delivery of large polypeptide molecules to the CNS, are likely to lead to the development of therapies aimed at enhancing inherent repair processes. Rodriguez and his colleagues at the Mayo Clinic in Rochester are making impressive strides in this respect (Van Engelen et al. 1994). Indeed, glial cell transplantation may find a role not only for the provision of myelinogenic cells but also as a means of either enhancing inherent remyelination (Franklin et al. 1991; Blakemore et al. 1995b) or introducing remyelination-enhancing agents into areas of demyelination.

Acknowledgements. The author holds a Wellcome Trust Research Career Development Fellowship and would like to thank Dr WF Blakemore for his immense contribution to much of the work described in the article.

References

Archer DR, Cuddon PA, Duncan ID (1992) Myelination by glial cells following transplantation into the CNS of the shaking pup. Rest Neurol Neurosci 4:151

Archer DR, Cuddon PA, Lipsitz D, Duncan ID (1997) Myelination of the canine central nervous system by glial cell transplantation: a model for repair of human myelin disease. Nature Medicine 3: 54–59

Archer DR, Leven S, Duncan ID (1994) Myelination by cryopreserved xenografts and allografts in the myelin-deficient rat. Exp Neurol 125:268–277

Armstrong RC, Dorn HH, Kufta CV, Friedman E, Dubois-Dalcq ME (1992) Pre-oligodendrocytes from adult human CNS. J Neurosci 12: 1538–1547

Barnett SC, Franklin RJM, Blakemore WF (1993a) In vitro and in vivo analysis of a rat bipotential O-2A progenitor cell line containing the temperature sensitive mutant gene of the SV 40 large T antigen. Eur J Neurosci 5:1247–1260

Barnett SC, Hutchins AM, Noble M (1993b) Purification of olfactory nerve ensheathing cells from the olfactory bulb. Dev Biol 155:337–350

Baron-Van Evercooren A, Clerin-Duhamel E, Lapie P, Gansmuller A, Lachapelle F, Gumpel M (1993) The fate of transplanted Schwann cells in the brain during development. Dev Neurosci 14:73–84

Baron-Van Evercooren A, Avellana-Adalid V, Ben Younes-Dhennoufi A, Gansmüller A, Nait-Oumesmar B, Vignais L (1996) Cell-cell interactions during the migration of myelin-forming cells transplanted in the demyelinated spinal cord. Glia 16:147–164

Barres BA, Raff MC (1994) Control of oligodendrocyte number in the developing rat optic nerve. Neuron 12: 935–942

Blakemore WF (1978) Invasion of Schwann cells into the spinal cord of the rat following local injections of lysolecithin. Neuropath Appl Neurobiol 4:47–59

Blakemore WF, Crang AJ (1985) The use of cultured autologous Schwann cells to remyelinate areas of persistent demyelination in the central nervous system. J Neurol Sci 70:207–223

Blakemore WF, Crang AJ (1988) Extensive oligodendrocyte remyelination following injection of cultured central nervous system cells into demyelinating lesions in adult central nervous system. Dev Neurosci 10:1–11

Blakemore WF, Franklin RJM (1991) Transplantation of glial cells in the CNS. Trends Neurosci 14:323–327

Blakemore WF, Crang AJ, Curtis R (1986) The interaction of Schwann cells with CNS axons in regions containing normal astrocytes. Acta Neuropathol 71:295–300

Blakemore WF, Crang AJ, Franklin RJM, Tang K, Ryder S (1995a) Glial cell transplants that are subsequently rejected can be used to influence regeneration of glial cell environments in the CNS. Glia 13:79–91

Blakemore WF, Crang AJ, Franklin RJM (1995b) Transplantation of glial cells. In: Kettenman H, Ransom B (eds) Neuroglia. Oxford University Press, Oxford, pp 869–882

Crang AJ, Blakemore WF (1991) Remyelination of demyelinated rat axons by transplanted mouse oligodendrocytes. Glia 4: 305–313

Doucette R (1990) Glial influences on axonal growth in the primary olfactory system. Glia 3:433–449

Duncan ID (1996) Glial cell transplatation and remyelination of the central nervous system. Neuropathol Appl Neurobiol 22:87–1000

Duncan ID, Hammang JP, Jackson KF, Wood PM, Bunge RP, Langford L (1988) Transplantation of oligodendrocytes, and Schwann cells into the spinal cord of the myelin deficent rat. J Neurocytol 17:351–360

Duncan ID, Paino C, Archer DR, Wood PM (1992) Functional capacities of transplanted cell-sorted adult oligodendrocytes. Dev Neurosci 14:114–122

Espinosa de los Monteros A, Zhang M, DeVellis J (1993) O-2A progenitor cells transplanted into the neonatal rat brain develop into oligodendrocytes but not astrocytes. Proc Natl Acad Sci USA 90:50–54

Felts PA, Smith KJ (1992) Conduction properties of central nerve fibres remyelinated by Schwann cells. Brain Res 574:178–192

Franklin RJM (1993) Reconstructing myelin-deficient environments in the CNS by glial cell transplantation. Semin Neurosci 5:443–451

Franklin RJM, Blakemore WF (1995) Glial cell transplantation and plasticity in the O-2A lineage – Implications for CNS repair. Trends Neurosci 18:151–156

Franklin RJM, ffrench-Constant C (1996) Transplantation and repair in multiple sclerosis. In: Russell WC (ed) The molecular biology of multiple sclerosis. Wiley, New York, pp 231–242

Franklin RJM, Crang AJ, Blakemore WF (1991) Transplanted type-1 astrocytes facilitate repair of demyelinating lesions by host oligodendrocytes in adult rat spinal cord. J Neurocytol 20:420–430

Franklin RJM, Crang AJ, Blakemore WF (1993) The reconstruction of an astrocytic environment in glia-deficient areas of white matter. J Neurocytol 22:382–394

Franklin RJM, Bayley SA, Blakemore WF (1996a) Transplanted CG4 cells (an oligodendrocyte progenitor cell line) survive, migrate and contribute to the repair of areas of demyelination in X-irradiated and damaged spinal cord, but not in normal spinal cord. Exp Neurol 137:263–276

Franklin RJM, Bayley SA, Milner R, ffrench-Constant C, Blakemore WF (1995) Differentiation of the O-2A progenitor cell line CG-4 into oligodendrocytes and astrocytes following transplantation into glia-defient areas of CNS white matter. Glia 13:39–44

Franklin RJM, Gilson JM, Franceschini IA, Barnett SC (1996) Schwann cell-like myelination following transplantation of an olfactory-bulb-ensheath-ing-cell line into areas of demyelination in the adult CNS. Glia 17:217–224

Friedrich VL, Lazzarini RA (1993) Restricted migration of transplanted oligodendrocyte or their progenitors, revealed by transgenic marker MBP. J Neural Transpl Plast 4:139–146

Gansmüller A, Clerin E, Krüger F, Gumpel M, Lachapelle F (1991) Tracing transplanted oligodendrocytes during migration and maturation in the shiverer mouse brain. Glia 4:588–590

Ghirnikar RS, Eng LF (1994) Astrocyte-Schwann cell interactions in culture. Glia 11:367–377

Gilson J, Blakemore WF (1993) Failure of remyelination in areas of demyelination produced in the spinal cord of old rats. Neuropath Appl Neurobiol 19:173–181

Gogate N, Verma L, Zhou JM, Milward E, Rusten R, O'Connor M, Kufta K, Kim J, Hudson L, Dubois-Dalcq M (1994) Plasticity in the adult human oligodendrocyte lineage. J Neurosci 14:4571–4587

Gout O, Dubois-Dalcq M (1993) Directed migration of transplanted glial cells toward a spinal cord demyelinating lesion. Int J Dev Neurosci 11:613–623

Groves AK, Barnett SC, Franklin RJM, Crang AJ, Mayer M, Blakemore WF, Noble M (1993) Repair of demyelinated lesions by transplantation of purified O-2A progenitor cells. Nature 362:453–455

Gumpel M, Gout O, Gansmüller A, Baumann N (1989) Myelination and remyelination in the central nervous system by transplanted oligodendrocytes using the shiverer model. Dev Neurosci 11:132–139

Honmou O, Felts PA, Waxman SG, Kocsis JD (1996) Restoration of normal conduction properties in demyelinated spinal cord axons in the adult rat by transplantion of exogenous Schwann cells. J Neurosci 16:3199–3208

Huang PP, Alliquant B, Carmel PW, Friedman ED (1991) Myelination of the rat retina by transplantation of oligodendrocytes into 4-day-old hosts. Exp Neurol 113:291–300

Jacque C, Quinonero J, Collins PV, Villarroya H, Suard I (1992) Comparative migration and development of astroglial and oligodendroglial cell populations from a brain xenograft. J Neurosci 12:3098–3106

Jeffery ND, Blakemore WF (1997) Locomotor deficits induced by experimental spinal cord demyelination are abolished by spontaneous remyelination. Brain 120: 27–37

Komoly S, Hudson LD, Webster H deF, Bondy CA (1992) Insulin-like growth factor gene expression is induced in astrocytes during experimental demyelination. Proc Natl Acad Sci USA 89:1894–1898

Lachapelle F, Gumpel M, Baulac M, Jacque C, Duc P, Baumann N (1984) Transplantation of CNS fragments into brain of shiverer mutant mice: extensive myelination by implanted oligodendrocytes. I. immunohistochemical studies. Dev Neurosci 6:325–334

Lachapelle F, Duhamel-Clerin E, Gansmüller A, Baron-Van Evercooren A, Villarroya H, Gumpel M (1994) Transplanted transgenically marked oligodendrocytes survive, migrate and myelinate in the normal mouse brain as they do in the shiverer mouse brain. Eur J Neurosci 6:814–824

Ludwin SK (1988) Remyelination in the central nervous system and the peripheral nervous system. In: Waxman SG (ed) Advances in neurology, Vol 47. Raven, New York, pp 215–254

Ludwin SK, Szuchet S (1993) Myelination by mature ovine oligodendrocytes in vivo and in vitro: evidence that different steps in the myelination process are independently controlled. Glia 8:219–231

McKay JS, Blakemore WF, Franklin RJM (1997) The effects of the growth-factor antagonist trapidil on remyelination in the CNS. Neuropath Appl Neurobiol 23: 50–58

O'Leary MT, Blakemore WF (1997) Oligodendrocyte precursors survive poorly and do not migrate following transplantation into normal adult central nervous system. J Neurosci Res 48:159–167

Oumesmar BN, Vignais L, Duhamel-Clerin E, Avellana-Adalid V, Rougon G, Baron-Van Evercooren A (1995) Expression of the highly polysialylated neural cell adhesion molecule during postnatal myelination and following chemically induced demyelination of the adult mouse spinal cord. Eur J Neurosci 7:480–491

Pfeiffer SE, Warrington AE, Bansal R (1993) The oligodendrocyte and its many cellular processes. Trends Cell Biol 3:191–197

Prineas JW, Barnard RO, Kwon EE, Sharer L, Cho ES (1993a) Multiple sclerosis: remyelination of nascent lesions. Ann Neurol 33:137–51

Prineas JW, Barnard RO, Revesz T, Kwon EE, Sharer L, Cho ES (1993b) Multiple sclerosis: pathology of recurrent lesions. Brain 116:681–93

Rosenbluth J, Hasegawa M, Shirasaki N, Rosen CL, Liu Z (1990) Myelin formation following transplantation of normal fetal glia into myelin deficient rat spinal cord. J Neurocytol 19:718–730

Rosenbluth J, Liu Z, Guo D, Schiff R (1993) Myelin formation by mouse glia in myelin-deficient rats treated with cyclosporine. J Neurocytol 22:967–977

Rutkowski JL, Kirk CJ, Lerner MA, Tennekoon GI (1995) Purification and expansion of human Schwann cells in vitro. Nature Med 1:80–83

Scolding N, Rayner PJ, Sussman J, Shaw C, Compston DAS (1995) A proliferative adult human oligodendrocyte progenitor. NeuroReport 6:441–445

Smith KJ, Blakemore WF, McDonald WI (1979) Central remyelination secures secure conduction. Nature 280:395–396

Targett MP, Sussman J, Scolding N, O'Leary MT, Compston DAS, Blakemore WF (1996) Failure to achieve remyelination of demyelinated axons following transplantation of glial cells obtained from adult human brain. Neuropath Appl Neurobiol 22:199–206

Tontsch U, Archer DR, Dubois-Dalcq M, Duncan ID (1994) Transplantation of an oligodendrocyte cell line leading to extensive myelination. Proc Natl Acad Sci USA 91:11616–11620

Trotter J, Crang AJ, Schachner M, Blakemore WF (1993) Lines of glial precursor cells immortalised with a temperature-sensitive oncogene give rise to astrocytes and myelin-forming oligodendrocytes on transplanation into demyelinated lesions in the central nervous system. Glia 9:25–40

Tourbah A, Baron-Van Evercooren A, Oliver L, Raulais D, Jeanny JC, Gumpel M (1992) Endogenous aFGF and cellular changes after a demyelinating lesion in the spinal cord of adult normal mice: immunohistichemical study. J Neurosci Res 33:47–59

Van Engelen BGM, Miller DJ, Pavelko KD, Hommes OR, Rodriguez M (1994) Promotion of remyelination by polyclonal immunoglobulin in Theiler's virus-induced demyelination and in multiple sclerosis. J Neurol Neurosurg Psychiatry 57 [Suppl]:65–68

Vignais L, Nait Oumesmar B, Mellouk F, Gout O, Labourdette G, Baron-Van Evercooren A, Gumpel M (1993) Transplantation of oligodendrocyte precursors in the adult demyelinated spinal cord: migration and remyelination. Int J Dev Neurosci 11:603–612

Warrington AE, Barbarese E, Pfeiffer SE (1993) Differential myelinogenic capacity of specific developmental stages of the oligodendrocyte lineage upon transplantation into hypomyelinating hosts. J Neurosci Res 34:1–13

11 Monitoring of Suicide Gene Therapy with Positron Emission Tomography

U. Haberkorn

11.1 Suicide Gene Therapy

Transfer and expression of suicide genes into malignant tumor cells represents an attractive approach for human gene therapy. Suicide genes typically code for non-mammalian enzymes which convert nontoxic prodrugs into highly toxic metabolites. Therefore, systemic administration of the nontoxic prodrug results in the production of the active drug at the tumor site. Our research center is concentrating mainly on two suicide systems: cytosine deaminase (CD) and herpes simplex virus thymidine kinase (HSV-tk).

Cytosine deaminase, which is expressed in yeast and bacteria but not in mammalian organisms, converts the antifungal agent 5-fluorocytosine (5-FC) into the highly toxic 5-fluorouracil (5-FU). In mammalian cells no anabolic pathway is known which leads to incorporation of 5-FC into the nucleic acid fraction. Therefore pharmacological effects (due to conversion of 5-FC to 5-FU by the intestinal microflora) are moderate and allow administration of high therapeutic doses (Scholer 1980; Polak et al. 1976). The distribution of 5-FC in mammalian tissues is uniform and neither tissue-specific accumulation nor appreciable binding to plasma proteins has been observed (Polak et al. 1976; Koechlin et al. 1966). 5-FU exerts its toxic effect by interfering with DNA and protein synthesis due to substitution of uracil by 5-FU in RNA and inhibition of thymidilate synthetase by 5-fluorodeoxy-uridine monophosphate resulting in impaired DNA biosynthesis (Myers 1981). Nishiyama et al. (1985) implanted CD-containing capsules into rat gliomas and subsequently treated the animals by systemic adminstration of 5-FC. They observed significant amounts of 5-FU in the tumors as well as a decrease in tumor growth rate and systemic cytotoxicity. This approach to local chemotherapy was expanded by Wallace et al. (1994) for application in patients with disseminated tumor disease. They used monoclonal antibody (mAb)-enzyme conjugates to achieve a selective activation of 5-FC thereby achieving a sevenfold higher level of 5-FU in the tumor after administration of mAb-CD and 5-FC than obtained by systemic administration of 5-FU.

Gene therapy with HSV-tk as suicide gene has been performed in a variety of tumor models both in vitro and in vivo (Chen et al. 1994; Borrelli et al. 1988; Barba et al. 1994; Moolten and Wells 1990; Caruso et al. 1993; Oldfield et al. 1993; Culver et al. 1992; Ram et al. 1993). In contrast to human thymidine kinase HSV-TK is less specific and also phosphorylates nucleoside analogs such as acyclovir and ganciclovir (GCV) to their monophosphate metabolites (Keller et al. 1981). These monophosphates are subsequently phosphorylated by cellular kinases to the di- and triphosphates. After integration of the metabolites into DNA, chain termination occurs followed by cell death.

Targeting the enzyme to the tumor site may be achieved by transfer and expression of the suicide gene using recombinant retroviral vectors. Encouraging results were initially obtained in rat gliomas using a retroviral vector system for transfer and expression of the HSV-tk gene (Culver et al. 1992; Ram et al. 1993). Recently, in vitro and in vivo studies have further demonstrated the potency of the CD suicide system. Tumor cells which had been infected with a retrovirus carrying the CD gene showed a strict correlation between 5-FC sensitivity and CD enzyme activity (Huber et al. 1993; Mullen et al. 1992, 1994). However, although not all of the tumor cells have to be infected to obtain a sufficient therapeutic response, repeated injections of the recombinant retroviruses may be necessary to reach a therapeutic level of enzyme activity in the tumor. Therefore a prerequisite for gene therapy using a suicide system is monitoring of suicide gene expression in the tumor: (1) to decide if repeated gene transductions of the tumor are necessary, and (2) to find a therapeutic window of maximum gene expression and consecutive prodrug administration (Haberkorn et al. 1996). Since 5-FC as well as GCV can be labeled with ^{18}F with sufficient in vivo stability (Visser et al. 1985; Monclus et al. 1995), positron emission tomography (PET) may be applied to assess the enzyme activity in vivo.

Gene therapy using suicide genes is performed in two steps: first, the tumor is infected with recombinant viruses to introduce the suicide enzyme into the cells. To obtain a sufficient level of enzyme activity in the tumor multiple infections may be necessary. Second, the nontoxic prodrug is administered systemically. For the planning and individualization of gene therapy the enzyme activity induced in the tumor has to be estimated in order to achieve a therapeutically sufficient enzyme level before administration of the prodrug. Moreover, the measurement of therapy effects on tumor metabolism may be useful for the prediction of therapy outcome at an early stage of the treatment. PET using tracers of tumor metabolism has been applied for the evaluation of treatment response in a variety of tumors and therapeutic regimens (Haberkorn et al. 1993, 1994a; Rozenthal et al. 1989; Bergstrom et al. 1987), indicating that these tracers provide useful parameters for the early assessment of therapeutic efficacy.

11.2 Monitoring of Gene Therapy by the Assessment of Metabolic Effects

After transfection of a rat hepatoma cell line with a retroviral vector containing the HSV-tk gene different clones were established by G418 selection (Haberkorn et al. 1997b). Thereafter, uptake measurements using fluorodeoxyglucose (FDG), 3-O-methylglucose, aminoisobutyric acid (AIB) and methionine were performed in the same cell lines in the presence of different concentrations of GCV. These experiments were done up to 48 h after the onset of therapy. During GCV treatment therapy the uptake of FDG and 3-O-methylglucose increases up to

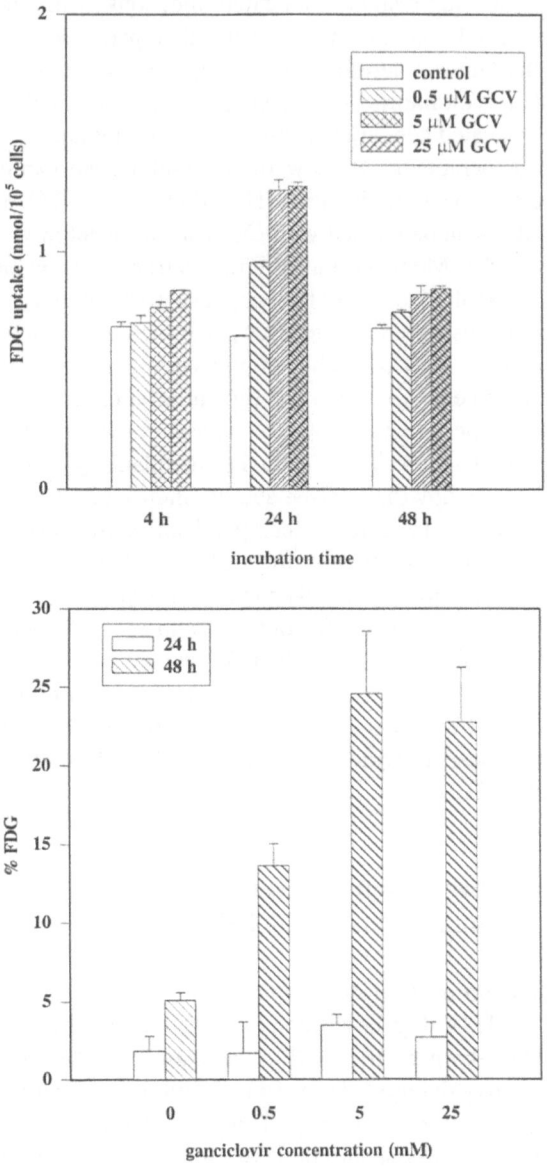

Fig. 1. Legend see p. 163

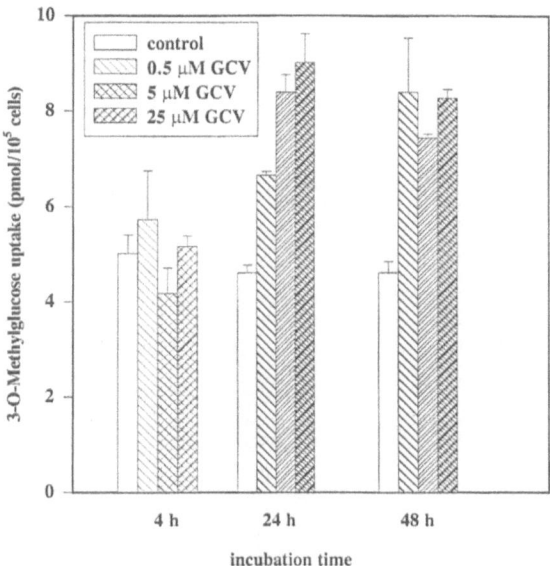

Fig. 2. 3-O-methylglucose uptake (pmol/10^5 cells) in HSV thymidine kinase (TK)-expressing cells after different incubation periods with 0.5, 5 and 25 mM ganciclovir (*GCV*). Mean and standard deviation (*n*=3)

195% after 24 h incubation with GCV (Figs. 1, 2). An HPLC analysis revealed a decline of the FDG-6-phosphate fraction after 48 h incubation with GCV (Fig. 1). Consequently, a normalization of FDG uptake was observed after this incubation period, whereas 3-O-methylglucose uptake was still increased. Experiments performed with different amounts of TK-expressing cells and control cells (bystander experiments) showed that these effects are dependent on the percentage of

◀ **Fig. 1.** Fluorodeoxyglucose (FDG) uptake (nmol/10^5 cells) in HSV thymidine kinase (TK)-expressing cells (**above**) after different incubation periods with 0.5, 5 and 25 mM ganciclovir (*GCV*). Content of FDG-6-phosphate (**below**) after 24 and 48 h exposure to the indicated concentrations of ganciclovir. Mean and standard deviation (*n*=3)

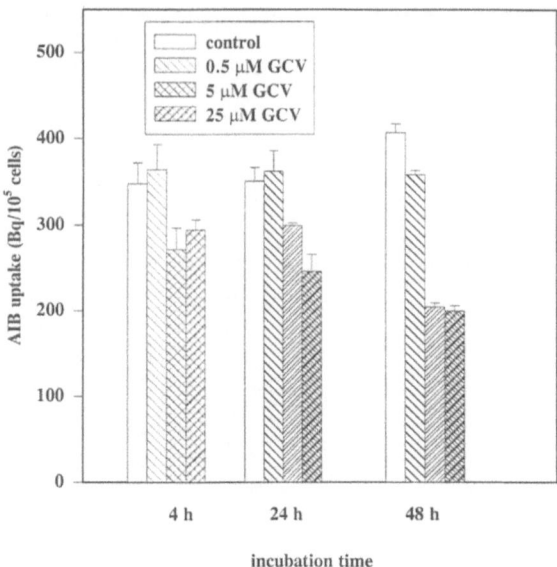

Fig. 3. Aminoisobutyric acid (AIB) uptake (Bq/10⁵ cells) in HSV thymidine kinase (TK)-expressing cells after different incubation periods with 0.5, 5 and 25 mM ganciclovir (*GCV*). Mean and standard deviation (*n*=3)

TK-expressing cells (Haberkorn et al. 1997a). The AIB uptake decreases to 47% (Fig. 3), while the methionine uptake in the acid-insoluble fraction decreases to 17% (Fig. 4).

In clinical and experimental studies an increase of FDG uptake early after treatment of malignant tumors was observed (Rozenthal et al. 1989; Haberkorn et al. 1992, 1994b). Cell culture experiments with rat adenocarcinoma cells under chemotherapy revealed that this effect is predominantly due to enhanced glucose transport (Haberkorn et al. 1994b). As an underlying mechanism, a redistribution of the glucose transport protein from intracellular pools to the plasma membrane may be considered. This has been observed in cell culture studies as a general reaction to cellular stress (Wertheimer et al. 1991; Widnell et al. 1990; Pasternak et al. 1991; Clancy and Czech 1990). Since prodrug activation by the HSV-TK leads to DNA chain termination and cell damage, the

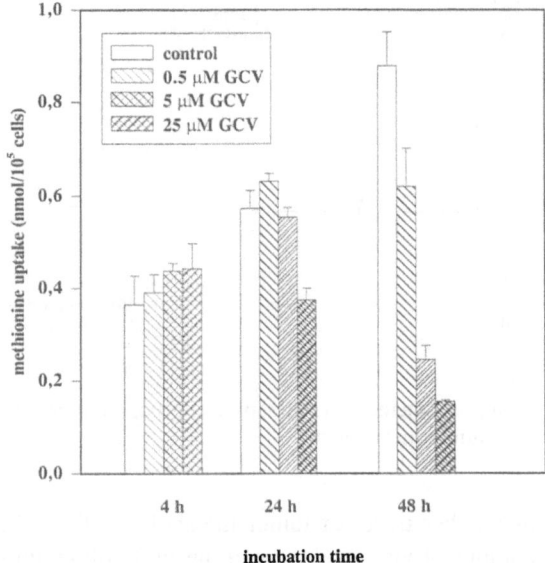

Fig. 4. Methionine uptake (nmol/10^5 cells) in the acid-insoluble fraction in HSV thymidine kinase (TK)-expressing cells after different incubation periods with 0.5, 5 and 25 mM ganciclovir (*GCV*). Mean and standard deviation (*n*=3)

same reactions may also occur in tumor cells under gene therapy with this suicide system. Translocation of glucose transport proteins to the plasma membrane as a first reaction to cellular stress may cause enhancement of glucose transport and represents a short-term regulatory mechanism which acts independent of protein synthesis. Therefore, the increase in FDG and 3-O-methylglucose uptake in vitro is interpreted as stress reaction of the tumor cells. However, an uncoupling of transport and phosphorylation was observed after 48 h incubation. The amino acid uptake experiments point to an inhibition of protein synthesis as well as of neutral amino acid transport.

These data indicate that combinations of the PET tracers used in these experiments may be used for monitoring gene therapy with HSV tk. Figure 5 shows how monitoring of suicide gene therapy with PET may be possible: first a baseline examination is performed with FDG,

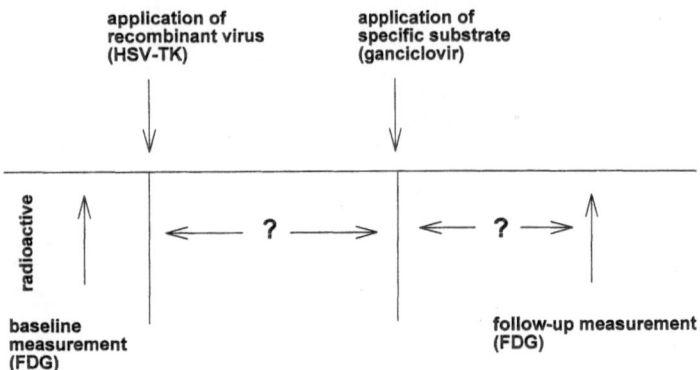

Fig. 5. Monitoring of suicide gene therapy using positron emission tomography and tracer of tumor metabolism

thymidine or another tracer of tumor metabolism. Then the infection with the recombinant virus is done and the PET follow-up studies are performed after administration of the prodrug. However, two problems remain: first, determination of the time period between the onset of prodrug administration and the follow-up PET studies. This question may be answered empirically. However, the other problem is more serious: how can we decide when a therapeutically sufficient suicide enzyme activity has been achieved in the tumor? This question can be adressed by measuring the uptake of specific substrates for the suicide systems.

11.3 Monitoring of Gene Therapy by the Uptake of Specific Substrates

In the same tumor system (Haberkorn et al. 1997b) uptake measurements were performed up to 48 h in a TK-expressing cell line and in a control cell line bearing the empty vector using thymidine (measured under therapy conditions), fluorodeoxycytidine (FdCyt) and GCV. In the TK-expressing cell line an increase (up to 250%) in thymidine uptake in the acid-soluble fraction and a decrease to 5.5% in the acid-insoluble fraction (Fig. 6) were found. The decreased radioactivity in the

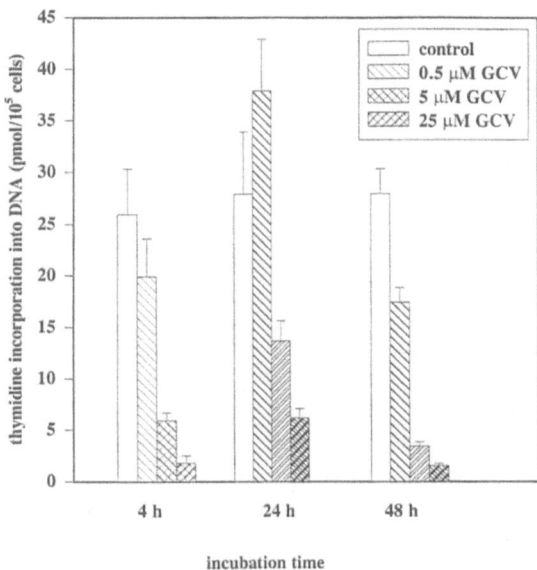

Fig. 6. Thymidine uptake (pmol/10^5 cells) in HSV thymidine kinase (TK)-expressing cells after different incubation periods with 0.5, 5 and 25 mM ganciclovir (*GCV*). Mean and standard deviation (*n*=3)

nucleic acid fraction occurs early (4 h) after exposure of the cells to GCV and represents DNA chain termination induced by the HSV-tk/GCV system. A posttherapeutic increase of TdR or its metabolites in the acid-soluble fraction was observed in previous studies after chemotherapy (Haberkorn et al. 1994a). This effect may be explained by an increase in the activity of salvage pathway enzymes, e.g., of host TK activity, during repair of cell damage. Therefore, it is unlikely that quantitation with a standardized uptake value is useful for determining proliferation in treated tumors. Correct interpretation of the PET signal (representing the total radioactivity in a given tissue volume) depends on information regarding TdR metabolism and the size of the different metabolite fractions in the tumor. Therefore, PET measurements with ^{11}C-TdR may be used to assess the effects of the HSV-tk/GCV system on DNA synthesis if quantitation is based on a modelling approach.

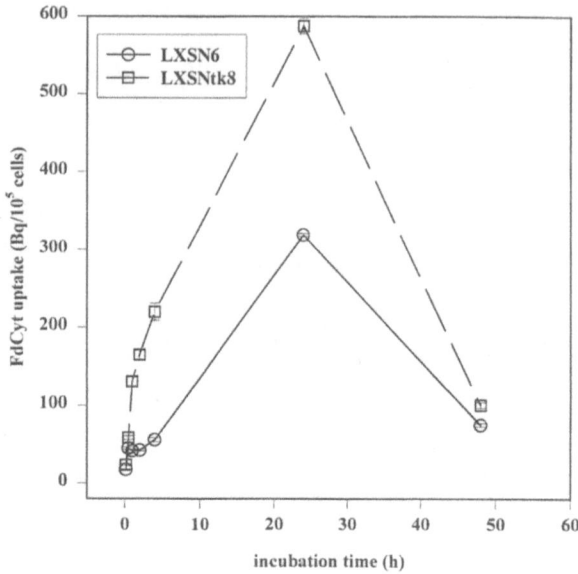

Fig. 7. Total fluorodeoxycytidine uptake (Bq/10⁵ cells) in HSV thymidine kinase (TK)-expressing cells *(LXSNtk8)* and control cells *(LXSN6)* after the indicated incubation times. Mean and standard deviations (*n*=3)

The FdCyt uptake was higher in the TK-expressing cells with a maximum after 4 h (12-fold and threefold higher in the acid-insoluble and acid-soluble fraction, respectively). After longer incubation periods the FdCyt uptake declined (Fig. 7). HPLC analysis showed a rapid and almost complete metabolism and degradation in both cell lines (Haberkorn et al. 1997b), which might be due to dehalogenation or the action of nucleosidases.

The GCV uptake showed a time-dependent increase in TK-expressing cells and a plateau in control cells (Fig. 8). The HPLC analysis revealed unmetabolized GCV in control cells and a time-dependent shift of GCV to its phosphorylated metabolite in TK-expressing cells (Haberkorn et al. 1997b). The difference in total GCV uptake between TK-expressing cells and control cells was higher (eightfold after 4 h, 30-fold after 48 h) than for FdCyt (three- to fourfold after 4 h, 1.5- to

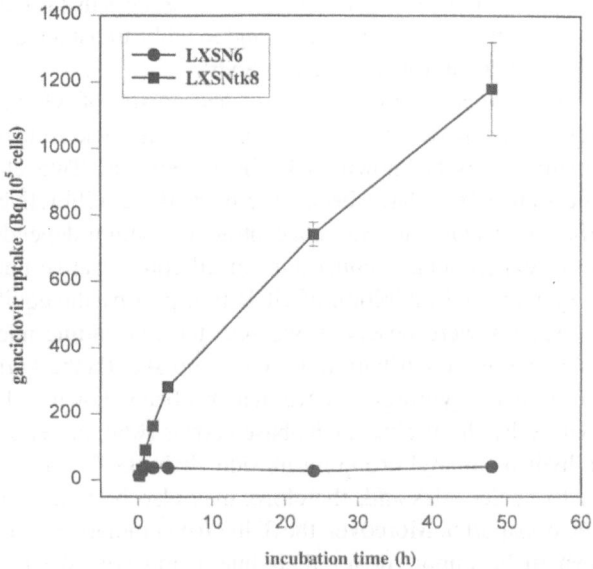

Fig. 8. Total ganciclovir uptake (Bq/10^5 cells) in HSV thymidine kinase (TK)-expressing cells (*LXSNtk8*) and control cells (*LXSN6*) after the indicated incubation times. Mean and standard deviations (*n=3*)

twofold after 48 h). Furthermore, the GCV uptake was highly correlated with the percentage of TK-expresing cells and with growth inhibition as measured in bystander experiments (Haberkorn et al. 1997b), i.e., in different mixtures of HSV-TK expressing cells and control cells. GCV can be labeled with ^{18}F and therefore is a candidate for PET measurements. Due to the greater difference in tracer uptake and the high amount of FdCyt degradation, the use of labeled GCV for quantitative PET studies seems to be superior to use of labeled FdCyt. In a recent study 5-iodo-2'-fluoro-2'deoxy-1-β-D-arabinofuranosyluracil (FIAU) was reported as a better marker substrate for TK activity (Tjuvajev et al. 1995). However, the authors related the FIAU uptake to the GCV sensitivity and the level of tk expression in bulk cultures of transduced cell lines or in different cell lines. Therefore, differences of the integration site and possible inactivation of the viral promoter may influence the FIAU uptake in this study. In our study differences in TK expression were simulated by the mixture of a single stably transfected cell line (LXSNtk8) with a control cell line (LXSN6).

To further elucidate the transport mechanism of GCV, inhibition/competition experiments were performed. The nucleoside transport in mammalian cells is known to be heterogeneous. Two classes of nucleoside transporters have been described: the equilibrative, facilitated diffusion systems and the concentrative, sodium-dependent systems. In our experiments competition for all concentrative nucleoside transport systems and inhibition of GCV transport by the equilibrative transport systems were observed, whereas the pyrimidine nucleobase system showed no contribution to GCV uptake (Haberkorn et al. 1997b). In human erythrocytes acyclovir has been shown to be transported mainly by the purine nucleobase carrier (Mahony et al. 1988). Due to a hydroxymethyl group on its side chain, GCV has a stronger similarity to nucleosides and, therefore, may also be transported by a nucleoside transporter. Moreover, the 3' hydroxyl moiety of nucleosides was shown to be important for their interaction with the nucleoside transporter (Gati et al. 1984).

In human glioblastoma cells the CD suicide system was evaluated. As mentioned above, CD-expressing cells convert the nontoxic prodrug 5-FC to the toxic metabolite 5- FU. PET with 5-(^{18}F)FC may be used for in vivo measurement of CD activity in genetically modified tumors. A human glioblastoma cell line was stably transfected with the *E. coli* CD gene (Haberkorn et al. 1996). After incubation of lysates of both CD-expressing cells and controls with (^{3}H)-5-FC HPLC analysis was done. The uptake of 5-FC was measured after different incubation times using therapeutic amounts of 5-FC. Saturation and competition experiments with 5-FC and 5-FU were performed and efflux was measured. ^{3}H -5-FU was produced in CD-expressing cells, whereas in the control cells only ^{3}H -5-FC was detected (Haberkorn et al. 1996). Moreover, significant amounts of 5-FU were found in the medium of cultured cells, which may account for the bystander effect observed in previous experiments. However, uptake studies revealed a moderate and nonsaturable

Fig. 9. 5-Fluorocytosine (*5-FC*) uptake (pmol/10^5 viable cells) in acid-insoluble (**above**) and acid-soluble fraction (**below**) of T1115RSV (controls) and T1115CD4 (CD-expressing) cells ▶

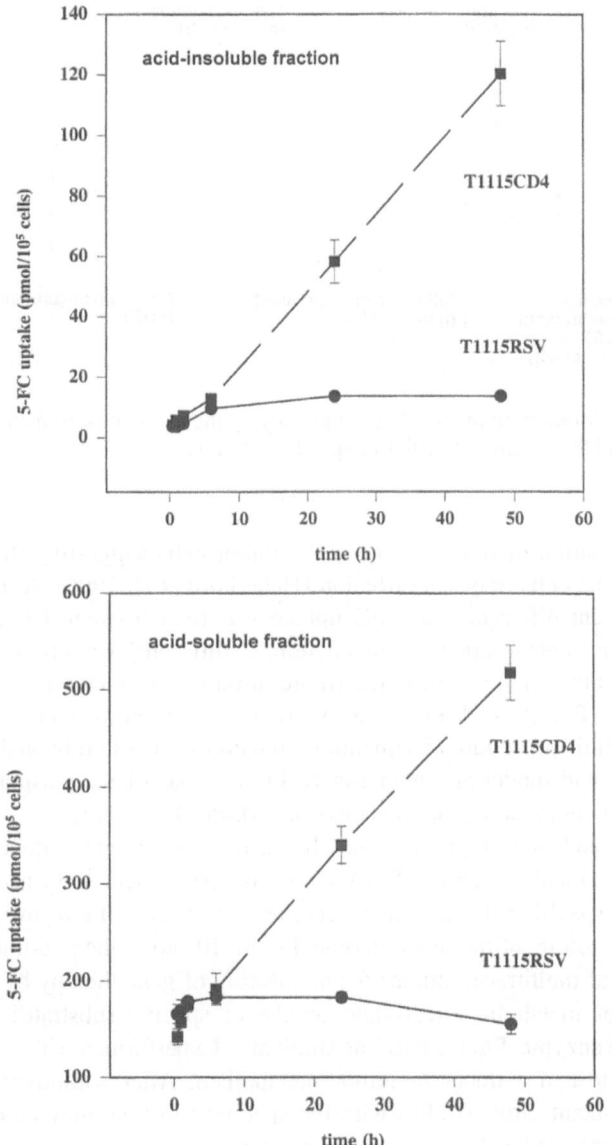

Fig. 9. Legend see p. 170

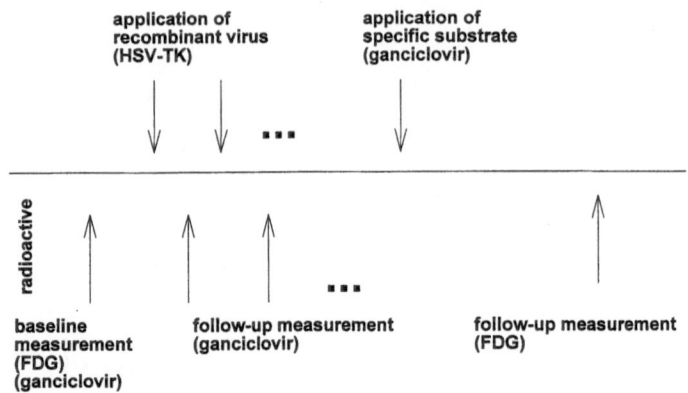

Fig. 10. Monitoring of suicide gene therapy by the assessment of metabolic effects and the pharmacokinetics of specific substrates

accumulation of radioactivity in the tumor cells suggesting that 5-FC enters the cells only via diffusion (Haberkorn et al. 1996). Although a significant difference in 5-FC uptake was seen between CD-positive cells and controls after 48 h incubation, no difference was observed after 2 h incubation (Fig. 9). Furthermore, a rapid efflux could be demonstrated. Therefore, 5-FC transport may be a limiting factor for this therapeutic procedure and quantitation with PET has to rely on dynamic studies and modeling, including HPLC analysis of the plasma, rather than on nonmodeling approaches (Haberkorn et al. 1996).

In conclusion, thymidine may be used to measure the effects of the HSV-tk suicide system on DNA synthesis. Assessment of the accumulation of specific substrates may serve as an indicator of enzyme activity and, therefore, of therapy outcome. Figure 10 shows the possible application of multitracer studies for monitoring of gene therapy by assessment of metabolic effects and uptake of specific substrates for the suicide enzyme. First, a baseline study may be performed with a specific substrate and a tracer of tumor metabolism. After administration of recombinant virus, studies with the specific substrate may be done to determine the level of suicide enzyme activity in the tumor. This information leads to the decision of whether the prodrug can be given. Thereafter, PET studies of tumor metabolism can be used for assessment of therapeutic efficacy. Since GCV is transported by the nucleoside carriers and may be trapped after phosphorylation, this suicide system seems to allow better monitoring of therapy than the 5-FC/CD system, as 5-FC enters the cell by passive diffusion and also shows rapid efflux (Haberkorn et al. 1996). However, animal experiments are needed to assess whether the in vivo signal obtained with PET is sufficient to obtain significant differences for determining a cut-off point between therapy responders and non-responders.

References

Barba D, Hardin J, Sadelain M, Gage FH (1994) Development of anti-tumor immunity following thymidine kinase-mediated killing of expermental brain tumors. Proc Natl Acad Sci USA 91:4348–4352

Bergstrom M, Muhr C, Lundberg PO, Bergstrom K, Gee AD, Fasth KJ, Langstrom B (1987) Rapid decrease in amino acid metabolism in prolactin-secreting pituitary adenomas after bromocriptine treatment: a PET study. J Comput Assist Tomogr 11:815–9

Borrelli E, Heyman R, Hsi M, Evans RM. Targeting of an inducible toxic phenotype in animal cells (1988) Proc Natl Acad Sci USA 85:7572–7576

Caruso M, Panis Y, Gagandeep S, Houssin D, Salzmann JL, Klatzman D (1993) Regression of established macroscopic liver metastases after in situ transduction of a suicide gene. Proc Natl Acad Sci USA 90:7024–7028

Clancy BM, Czech MP (1990) Hexose transport stimulation and membrane redistribution of glucose transporter isoforms in response to cholera toxin, dibutyryl cyclic AMP, and insulin in 3T3 adipocytes. J Biol Chem 265:12434–12443

Chen SH, Shine HD, Goodman JC, Grossman RG, Woo SLC (1988) Gene therapy for brain tumors: regression of experimental gliomas by adenovirus-mediated gene transfer in vivo. Proc Natl Acad Sci USA 91:3054–3057

Culver KW, Ram Z, Walbridge S et al. (1992) In vivo gene transfer with retroviral vector-producer cells for treatment of experimental brain tumors. Science 256:1550–1552

Gati WP, Misra HK, Knaus EE, Wiebe LI (1984) Structural modifications at the 2' and 3' positions of some pyrimidine nucleosides as determnants of their interaction with the mouse erythrocyte nucleoside transporter Biochem Pharmacol 33:3325–3331

Haberkorn U, Reinhardt M, Strauss LG, et al. (1992) Metabolic design of combination therapy: Use of enhanced fluorodeoxyglucose uptake caused by chemotherapy. J Nucl Med 33:1981–1987

Haberkorn U, Strauss LG, Dimitrakopoulou A, et al. (1993) Fluorodeoxyglucose imaging of advanced head and neck cancer after chemotherapy. J Nucl Med 34:12–17

Haberkorn U, Oberdorfer F, Klenner T et al. (1994a) Metabolic and transcriptional changes in osteosarcoma cells treated with chemotherapeutic drugs. Nucl Med Biol 21:835–845

Haberkorn U, Morr I, Oberdorfer F, et al. (1994b) Fluorodeoxyglucose uptake in vitro: aspects of method and effects of treatment with gemcitabine. J Nucl Med 35:1842–1850

Haberkorn U, Oberdorfer F, Gebert J et al. (1996) Monitoring of gene therapy with cytosine deaminase: in vitro studies using ^{3}H-5-fluorocytosine. J Nucl Med 37:87–94

Haberkorn U, Altmann A, Morr I. (1997a) Multi tracer studies during gene therapy of hepatoma cells with HSV thymidine kinase and ganciclovir. J Nucl Med 38 (in press)

Haberkorn U, Altmann A, Morr I, et al. (1997b) Gene therapy with herpes simplex virus thymidine kinase in hepatoma cells: uptake of specific substrates. J Nucl Med 38: 287-294

Huber BE, Austin EA, Good SS, Knick VC, Tibbels S, Richards CA (1993) In vivo antitumor activity of 5-fluorocytosine on human colorectal carcinoma cells genetically modified to express cytosine deaminase. Cancer Res 53:4619–4626

Keller PM, Fyfe JA, Beauchamp L et al. (1981) Enzymatic phosphorylation of acyclic nucleoside analogs and correlations with antiherpetic activities. Biochem Pharmacol 30:3071–3077

Koechlin BA, Rubio F, Palmer S, Gabriel T, Duschinsky R (1966) The metabolism of 5-fluorocytosine-2-^{14}C and of cytosine-1-^{14}C in the rat and the disposition of 5-fluorocytosine-2-^{14}C in man. Biochem Pharmac 15:435–446

Mahony WB, Domin BA, McConnel RT, Zimmerman TP (1988) Acyclovir transport into human erythrocytes. J Biol Chem 263:9285–9291

Monclus M, Luxen A, Van Naemen J et al. (1995) Development of PET radiopharmaceuticals for gene therapy: synthesis of 9-((1-(^{18}F)fluoro-3-hydroxy-2-propoxy)methyl)guanine. J Label Comp Radiopharm 37:193–195

Moolten FL, Wells JM. Curability of tumors bearing herpes thymidine kinase genes transferred by retroviral vectors. J Natl Cancer Inst 82:297–300

Mullen CA, Kilstrup M, Blaese M (1992) Transfer of the bacterial gene for cytosine deaminase to mammalian cells confers lethal sensitivity to 5-fluorocytosine: a negative selection system. Proc Natl Acad Sci USA 89:33–37

Mullen CA, Coale MM, Lowe R, Blaese RM (1994) Tumors expressing the cytosine deaminase suicide gene can be eliminated in vivo with 5-fluorocytosine and induce protective immunity to wild type tumor. Cancer Res 54:1503–1506

Myers CE (1981) The pharmacology of the fluoropyrimidines. Pharmacol Rev 33:1–15

Nishiyama T, Kawamura Y, Kawamoto K, Matsumura H, Yamamoto N, Ito T, Ohyama A, Katsuragi T, Sakai T (1985) Antineoplastic effects of 5-fluorocytosine in combination with cytosine deaminase capsules. Cancer Res 45:1753–1761

Oldfield EH, Ram Z, Culver KW, Blaese RM, DeVroom HL, Anderson WF (1993) Gene therapy for the treatment of brain tumors using intra-tumoral transduction with the thymidine kinase gene and intravenous ganciclovir. Hum Gene Ther 1:39–69

Pasternak CA, Aiyathurai JEJ, Makinde V, et al. (1991) Regulation of glucose uptake by stressed cells. J Cell Physiol 149:324–331

Polak A, Eschenhof E, Fernex M, Scholer HJ (1976) Metabolic studies with 5-fluorocytosine-6-^{14}C in mouse, rat, rabbit, dog and man. Chemotherapy 22:137–153

Ram Z, Culver WK, Walbridge S et al. (1993) In situ retroviral-mediated gene transfer for the treatment of brain tumors in rats. Cancer Res 53:83–33

Rozenthal JM, Levine RL, Nickles RJ, Dobkin JA (1989) Glucose uptake by gliomas after treatment. Arch Neurol 46:1302–1307

Scholer HJ (1980) Flucytosine. In: Speller DCE (ed) Antifungal chemotherapy. John Wiley & Sons, Chichester New York, pp 35–106

Tjuvajev JG, Stockhammer G, Desai R, Uehara H, Watanabe K, Gansbacher B, Blasberg RG (1995) Imaging the expression of transfected genes in vivo. Cancer Res 55:6126–6132

Visser GWM, Boele S, Knops GHJN, Herscheid JDM, Hoekstra A (1985) Synthesis and biodistribution of (^{18}F)-5-fluorocytosine. Nucl Med Comm 6:455–459

Wallace PM, MacMaster JF, Smith VF, Kerr DE, Senter PD, Cosand WL (1994) Intratumoral generation of 5-fluorouracil mediated by an antibody-cytosine deaminase conjugate in combination with 5-fluorocytosine. Cancer Res 54:2719–2723

Wertheimer E, Sasson S, Cerasi E, Ben-Neriah Y (1991) The ubiquitous glucose transporter GLUT-1 belongs to the glucose-regulated protein family of stress-inducible proteins. Proc Natl Acad Sci USA 88:2525–2529

Widnell CC, Baldwin SA, Davies A, Martin S, Pasternak CA (1990) Cellular stress induces a redistribution of the glucose transporter. FASEB J 4:1634–1637

Subject Index

Ernst Schering Research Foundation Workshop

Editors: Günter Stock
Ursula-F. Habenicht